W9-AQE-418

VISIT TO A SMALL UNIVERSE

Masters of Modern Physics

Advisory Board

Dale Corson, Cornell University
Samuel Devons, Columbia University
Sidney Drell, Stanford Linear Accelerator Center
Herman Feshbach, Massachusetts Institute of Technology
Marvin Goldberger, Institute for Advanced Study, Princeton
Wolfgang Panofsky, Stanford Linear Accelerator Center
William Press, Harvard University

Series Editor

Robert N. Ubell

Published Volumes

The Road from Los Alamos by Hans A. Bethe
The Charm of Physics by Sheldon L. Glashow
Citizen Scientist by Frank von Hippel

VISIT TO A SMALL UNIVERSE

UNIVERSE

VIRGINIA TRIMBLE

The American Institute of Physics

ST. PHILIP'S COLLEGE LIBRARY

© 1992 by American Institute of Physics.
All rights reserved.
Printed in the United States of America.

Copyright and permissions notices for use of previously published material are
provided in the Acknowledgments section at the back of this volume.

American Institute of Physics
335 East 45th Street
New York, NY 10017-3483

Library of Congress Cataloging-in-Publication Data

Trimble, Virginia.
    Visit to a small universe / Virginia Trimble.
        p.   cm. – (Masters of modern physics)
    Includes index.
    ISBN 0–88318–792–2
    1. Astronomy.   I. Title   II. Series
QB51.T75 1991
520—dc20   91-42614
                                                                               CIP

This book is volume four of the Masters of Modern Physics series.

Contents

About the Series   viii

Preface   ix

ASTRONOMY, ANCIENT AND MODERN

Cheops' Pyramid   3

Star of Bethlehem   9

In Their Own Words   20

Relativistic Astrophysics   28

ORIGINS

According to Hoyle   49

Are Galaxies Here to Stay?   52
    with Martin Rees

Man's Place in the Universe   60

Dark Matter   84

The Anthropic Principle   94

It's a Nice Planet to Visit, but I Wouldn't Want to Live There   99

Where Are They?   105

LIVES OF STARS

Classifying Ourselves   115

The Odd Two Percent   119

Close Binary Stars   126

Cataclysmic Variables   137

DEATHS OF STARS

White Dwarfs: The Once and Future Suns   147

The Crab Nebula and Pulsar   158

The Greatest Supernova Since Kepler   167

So You Want to Find a Supernova?   187

Remnants of Historical Supernovae   195
    with David H. Clark

FRUITS OF MARRIAGE

Gravitational Radiation and the Binary Pulsar   207

Masada, Suicide, and Halakhah   210

Gravitational Radiation Detectors   221

ASTRONOMERS OBSERVED

The Best Is Yet to Be   235

Information Explosion   238

Progress Is Not Our Most Important Product   242

Death Comes as the End   244

Young Versus Established American Astronomers   250

Prestigious Start   261

Self-Citation Rates in Astronomical Papers   265

ANTIC MUSE

Brief Encounter with a Facilities Manual   271

Astronomical Conferences   273

Trimble's Laws   276

STAR GAZERS

Martin J. Rees   279

Maarten Schmidt   282

Beatrice M. Tinsley   285

Fritz Zwicky   294

References   301

Acknowledgments   323

Index   329

About the Author   335

About the Series

Masters of Modern Physics introduces the work and thought of some of the most celebrated physicists of our day. These collected essays offer a panoramic tour of the way science works, how it affects our lives, and what it means to those who practice it. Authors report from the horizons of modern research, provide engaging sketches of friends and colleagues, and reflect on the social, economic, and political consequences of the scientific and technical enterprise.

Authors have been selected for their contributions to science and for their keen ability to communicate to the general reader—often with wit, frequently in fine literary style. All have been honored by their peers and most have been prominent in shaping debates in science, technology, and public policy. Some have achieved distinction in social and cultural spheres outside the laboratory.

Many essays are drawn from popular and scientific magazines, newspapers, and journals. Still others—written for the series or drawn from notes for other occasions—appear for the first time. Authors have provided introductions and, where appropriate, annotations. Once selected for inclusion, the essays are carefully edited and updated so that each volume emerges as a finely shaped work.

Masters of Modern Physics is edited by Robert N. Ubell and overseen by an advisory panel of distinguished physicists. Sponsored by the American Institute of Physics, a consortium of major physics societies, the series serves as an authoritative survey of the people and ideas that have shaped twentieth-century science and society.

Preface

T he 36 pieces collected here are not precisely essays, at least as defined by my high school English teacher, in that the original version of each was written with some definite, not purely literary, purpose in mind. Come to think of it, so were our high school essays. (Their purpose was a B+.)

Of the items in this book, a few present results of my own investigations in areas outside mainstream astronomy. Many were intended to give wider exposure to exciting work done by colleagues than is possible through publication in research journals alone. Some are frankly didactic — rightly or wrongly, I really believe everyone should understand how stars have made the atoms in our bodies. Other pieces were requested by the organizers of conferences or by editors of books honoring distinguished scholars. And a couple (dare I confess?) were written mostly for money. When we were in graduate school, the Fundamental Monetary Unit (FMU) was the pitcher of beer at Shakey's ($1.35, if memory serves). For the academic scientist it is "the modest honorarium." Whether one of the fundamental units sounds like a little or a lot depends mostly on how many of them you happen to have floating around at a given time.

Astronomy, Ancient and Modern

Cheops' Pyramid is not the oldest object mentioned in this volume; the universe has it beat by 10 or 20 billion years. But it is the oldest writing, completed shortly before my 20th birthday, as a byproduct of a course at UCLA on Ancient Egyptian art and architecture. The professor, the late Alexandre Mikhail Badawy, learning that I was an astronomy major, suggested that I find out which stars had passed over the narrow shafts into the pyramid burial chamber during the years when it might have been constructed. The possible stars change with time because of precession of the equinoxes, and Badawy had in mind both verification of his hypothesis that the shafts had astrological significance and determination of the age of the pyramid independent of standard archeological methods. I have not kept up with the field, but I suspect the idea is no more popular now than it was then, and that other methods of dating are much more reliable.

Two things happen each year when the Christmas decorations begin to appear in the stores. One is Uncle Roy's birthday (30 September). The other is the first phone call asking what the star of Bethlehem was. At the time I wrote the book review whose modified version appears here, I was not convinced that the question could have a definite answer, or even that the question can be asked unambiguously. The intervening years have seen a good many more articles and lectures on the subject, but, so far as I know, no persuasive new ideas or better ways of deciding among the old ones. And Uncle Roy (who is an amateur astronomer) is still very difficult to shop for.

The other two items describe much more nearly contemporary events. Relativistic astrophysics acquired its name and status as a subdiscipline in 1963 (the year I wrote "Cheops' Pyramid"); and even the cosmological investigations described "in their own words" fall entirely within the lifetimes of astronomers who are still active in the field. Both pieces were written within the last year, and it is probable that, in a decade or two, different aspects of the history will be regarded as important.

Origins

Figuring out where each of us and people in general belong in the great scheme of things was once the province of religion and philosophy. But astronomy has been nibbling away at the edges at least since the time of Copernicus. We know roughly how old the universe is and something about its size and large scale structure. The earth and sun can be compared with other planets and stars and seen to be special in some ways, ordinary in others. The basic scheme has not changed much in the 15-year span over which these pieces were first written. It leads from a hot, dense early universe, to galaxies that form stars where nuclear reactions transform the simpler, lighter elements into the heavier ones needed by chemically-based life, on to planets whose stable environments permit energy from stars to interact with molecules of gradually increasing complexity, and finally to self-replicating (living) molecules, intelligence, and the ability to modify the home planet almost beyond recognition. Some parts of the scenario still need a good deal of work done on them.

Forming galaxies looked tricky 15 years ago, and it still does—more so, if anything. "Are Galaxies Here to Stay?" suggests that the early stages involved a spheroidal halo of low-mass stars. Halo first is still the most popular scheme, but recent models make those halos out of something other than normal stars, as described in "Dark Matter." We still have no firm evidence for the existence of planets around stars other than the sun, or for life other than on earth. If we did, you would have heard about it without reading this book, I promise.

The ideas in these pieces are, even more than in the other sections, an integration of everything I have ever been told or read. All too often, facts get folded into one's general world-view and their sources forgotten. William C. Saslaw (University of Virginia and Jesus College, Cambridge) first introduced me to the anthropic principle and Dale Russell (National Museum of Canada) to stenonychosaurus (now correctly known by another, taxonomically prior, name). Martin J. Rees (Cambridge University) appears explicitly as co-author of only one item ("Are Galaxies Here to Stay?"), but all of them have been strongly influenced both by his expert opinions on the universe and its contents and by his attitude of continuous reassessment of yesterday's ideas in light of today's data. Maybe some day we'll finish that book we started in 1978.

Lives of Stars

Stellar structure and evolution is in better shape than most other parts of modern astrophysics. A well-defined set of coupled differential equations incorporates the dominant physical processes in stars. And these equations can be solved (given enough patience, computer time, and laboratory data on atoms and nuclei) to calculate how stars of different masses, ages, and chemical compositions should appear. The surprise is that calculated stars look a lot like real ones. They have about the right sizes, brightnesses, temperatures, time scales, and relationships to each other. Star formation remains the most uncertain phase, and the mechanisms by which a single, massive gas cloud transforms itself into many stars are being constantly rethought. Mercifully, I have said very little about star formation in these pieces and so have relatively few words to eat. For both single stars and pairs, the full range of possible interactions and objects is even larger and more complicated than as described in "The Odd Two Percent" and "Close Binary Stars."

My first exposure to ideas of stellar structure and evolution came in two unusually good courses, taught by Thornton Leigh Page at UCLA and by J. Beverley Oke at Caltech. Since then, a very great number of colleagues, through their papers and lectures, have attempted to educate me further on the topic, above all Bohdan Paczyński (Warsaw and Princeton Universities).

Deaths of Stars

In general, ideas about specific astronomical objects and processes evolve rapidly. But I think it is fair to say that the major questions raised in these pieces, written between 1972 and 1990, are still with us. How many supernovae explode per century per galaxy, from which stars, with what energy sources, leading to what sorts of remnants, where should you look for them, and so forth? I would not currently disown any of the main answers suggested either. All are, however, subject to change with very little notice.

The Crab Nebula, which pops up several places in the section, was the subject of my Ph.D. dissertation in 1968. Thus my views were initially much influenced by my adviser, Guido Münch (now at Calar Alto Observatory and the Institute of Astronomy, Heidelberg), who was, in turn, the student of S. Chandrasekhar. More recently I have borrowed repeatedly and gratefully from research on stellar deaths by W. David Arnett (University of Arizona), Sidney van den Bergh (Dominion Astrophysical Observatory, British Columbia), Jesse L. Greenstein (California Institute of Technology), Ken'ichi Nomoto (University of Tokyo), J. Craig Wheeler (University of Texas), Lodewijk Woltjer (Haute Provence Observatory), and Stanford E. Woosley (University of California, Santa Cruz).

Fruits of Marriage

Dorothy L. Sayers has Harriet Vane (in *Gaudy Night*, 1937) say that women are likely to marry into their jobs. Some things have changed in 50 years; but it is still true that more than half the married American women astronomers have spouses who are in astronomy or a closely related science. I am average in this as in many other things, and my marriage to physicist Joe Weber has brought me into closer contact with the field of gravitational radiation than would otherwise have been the case. It also partially motivated my conversion to Judaism.

The two gravitational radiation reports here are conventional pieces, apart from my closeness to the source material (and it was generous of the editors of *Sky and Telescope* and *Nature* to let me be the one to report on the topics). The Masada piece is an experiment. It began as a term paper for an undergraduate course, taught by Rabbi Meyer Greenberg, which I audited at the University of Maryland. The subject is a traditional one in Jewish law, interest in which revives with each resumption of war in the Middle East. But the style is that of an astrophysics review article. The combination worked well enough to be accepted by a standard journal (*Conservative Judaism*), but I have not tried it again. Richard Feynman used to say that it was part of the arrogance of physicists to believe that we could learn to do anything that anybody else can do. The belief is obviously truer for some physicists than for others, but I do see what he meant.

In the past few years, the binary pulsar has continued to behave exactly as expected. Measurement of the various orbit parameters has improved so that the masses of the neutron stars are known to better than 0.1 percent, and other relativistic effects are beginning to show up. Detectors for gravitational radiation have progressed less happily. The Maryland group no longer receives any federal funding for the project and simply keeps a room-temperature antenna operating more or less continuously. The various low-temperature facilities in other places are not collecting data most of the time; and construction of new detectors in the United States and elsewhere has been postponed for at least a year or two.

Astronomers Observed

Snooping is a very widespread human activity, if not a highly regarded one. In these pieces, I am more or less spying on my colleagues while they carry out and publish their research. The tone is generally light, but there is an underlying more somber message. It is that some aspects, at least, of scientific research are going to be more difficult, less pleasant, and therefore probably less productive in the near future than they have been in the recent past. It shows up in all sort of numbers—how many applications a postdoc has to submit to find a suitable job; the growing stacks of paper that must be read to keep up with your own field and the surrounding bureaucracy; hours per week devoted to committee meetings; how long it takes to get a paper refereed, accepted, and published; the plethora of proposals that have to be written before one gets funded; and so forth. None of these has improved since the pieces were written.

The people whose ideas and questions have entered into this section include Helmut Abt (Kitt Peak National Observatory and editor of the *Astrophysical Journal*), Robert Becker (University of California, Davis), Sir William McCrea (University of Sussex), and Lodewijk Woltjer (former director of the European Southern Observatory).

Antic Muse

Apart from anything else, scientific research can be rip-roaring good fun, an aspect of the activity that is remarkably difficult to convey to those who have not experienced it. (And I think nothing I have written does this at all well.) When a new experiment or calculation comes out right, one shares a bit of the triumphal exultation common to all creators. When things go wrong, there is at any rate the consolation of communal suffering. Description of my most recent failure (". . . and, as a result of a small error in setting up the observing files, I am now the world's foremost expert on molecular gas ten minutes of arc south of the Crab Nebula") has elicited a similar anecdote from every astronomer I've told it to. These three short items are all more or less from the "better luck next time" side of the scientific experience.

Star Gazers

Too often, one gets to say nice things about one's colleagues only after the fact. I am, therefore, grateful to Stephen Maran, editor of *IAU Today* (the newspaper of the 1988 Baltimore General Assembly of the International Astronomical Union) for the opportunity to talk about Martin Rees and Maarten Schmidt and have them correct the details. The other two biographical sketches were produced under more conventional circumstances. Beatrice Tinsley was so bright and so quick that she frightened many of her associates out of our integrals; we still miss her very much. I knew Fritz Zwicky only during the last decade of his remarkably varied career. It was not enough.

ASTRONOMY, ANCIENT AND MODERN

Cheops' Pyramid

The pyramid of Cheops at Giza is unique among the monuments of Egypt in several ways. Not only is it the largest, best built, and most thoroughly surveyed of the pyramids, but it possesses several architectural features not found elsewhere. Among the most obvious of these are two shafts leading north and south out of the King's Chamber and slanting up to open on opposite faces of the monument (Figure 1). Although the northern shaft makes an average angle with the horizontal of about $31°$ and the southern one an angle of $44.5°$, because the King's Chamber is located south of the vertical axis from the apex of the pyramid, the two shafts open nearly at the same height on the northern and southern faces.[1]

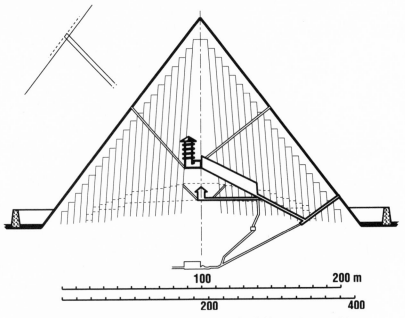

FIGURE 1. *Vertical cross-section north-south of Cheops' pyramid.*

ST. PHILIP'S COLLEGE LIBRARY

The purpose of these shafts has not been determined, but it has frequently been held that they were intended simply for ventilation, hence the name "air shafts." In view, however, of the profoundly religious character of the pyramids themselves for the ancient Egyptians, it seems not unreasonable to look for some deeper meaning to the shafts. Consider briefly some of the evidence for the view that the shafts were intended as ways whereby the soul of the deceased king might ascend to the north circumpolar stars and to the constellation now known as Orion. Although similar shafts do not appear to exist elsewhere, there is ample evidence for the presence of slots and apertures intended to allow the soul of the deceased to pass through various walls. Such apertures first appear in the Third Dynasty tomb of Djeser[2] and become a regular feature in the serdabs of the Fifth Dynasty mastaba tombs.[3]

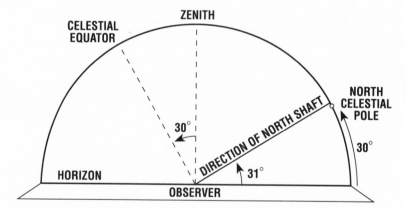

FIGURE 2. *Direction of north shaft and position of pole in sky for latitude 30° north.*

A notable feature of the religion of early Egypt was the "stellar destiny" of the soul, wherein it was thought that the soul of the dead king would rise to join the circumpolar stars—"The Indestructible Ones" or "The Imperishables" to the Egyptians—in their eternal journey around the sky. It is believed that the stairways or ramps descending from the north in archaic mastaba tombs were intended to aid the soul in its ascent to these stars. That the north shaft of the pyramid might have served a similar purpose is made more probable by its inclination. The latitude of Giza is about 30° north (29°58'51"), and we recall that the north shaft makes an angle of 31° with the horizontal. This means that the shaft points very nearly toward the north celestial pole, about which the circumpolar stars seem to revolve (Figure 2). It is also of interest to note that at the time the pyramid was built, the pole was marked by a bright star about as accurately as Polaris (α Ursae Minoris) now marks it.

It is generally known that the inclination of the earth's axis of rotation to the plane of its orbit (ecliptic) at an angle of about 23.5° combines with the nonspherical shape of the earth and the gravitational force of the sun, moon, and planets to produce a phenomenon known as the precession of the equinoxes. The effect of the sun and moon is to change the direction to which the earth's axis of rotation points relative to the fixed stars, while that of the planets is to change the plane of the earth's orbit relative to these stars. These effects are known as lunisolar precession and planetary precession respectively. It is evident that both factors will change the identity and positions of stars visible from a given point on the earth and that we must take them both into account when determining how the sky looked to the ancients.

In this scheme of moving stars, pole stars are a rather rare occurrence. In fact, after Polaris ceases to mark the pole in a few hundred years, there will not be another good one until α Draconis returns around A.D. 23,000.[*] It happens, however, that the last "visit" of α Draconis to the neighborhood of the pole occurred from about 3000 to 2500 B.C.[4] This means that the Egyptians of the pyramid age were more aware than might otherwise have been the case of the apparent daily journey of the stars about a fixed point in the sky. It thus seems highly probable that they would have chosen to build a shaft that would allow the soul of their dead king to ascend directly to this central point.

Noncircumpolar stars were also of considerable importance to the Egyptians. They measured time at night by means of decans—stars or groups of stars which rose or culminated (reached their highest elevation above the southern horizon) at one-hour intervals during the night. Many of these decans were parts of constellation pictures (although different from ours, which are derived from the Babylonian ones) and were identified with various gods. Very few of these have been identified with particular stars with any degree of certainty. There are, however, 4 of the 36 standard decans and 5 variants thereof which are parts of the constella-

---

[*]Although α Cephei and α Lyrae (Vega) will come within about 4° of the pole in A.D. 7500 and 14,000, they will not be nearly as accurate pole stars as Polaris and α Draconis, whose closest approaches to the pole are only about 30' away.

tion *Ś3ḥ*—"the god who crosses the sky"—whose identification with Orion "must be taken as likely in the highest degree."[5] He is depicted as a man standing, looking back over his shoulder and holding a scepter in one hand and an *'nḥ* sign in the other. One of the five variants is probably the "belt" of *Ś3ḥ*. Three of the decans intended for use during the epagomenal days appear also to have been parts of this constellation.[6] We may note as evidence for the identification the ceiling of the tomb of Senmut, in which the column devoted to *Ś3ḥ* includes three large stars arranged vertically and bearing a striking resemblance to the three stars we call Orion's belt (δ, ε, and ζ Orionis), which they probably represent.[7]

The next relevant question is, of course, the position of these stars relative to the southern shaft at the time the pyramid was built. This requires calculations to allow for the two types of precession previously noted. We observe first that, because the shaft is directed due south, it can point only to a star at culmination, and we see that for latitude 30° north and the inclination of the shaft, 44.5°, an appropriate star must have a declination (angular distance from the celestial equator) of –15.5° (Figure 3). The question is then reduced to whether or not the stars of Orion ever had such a declination and, if so, when.

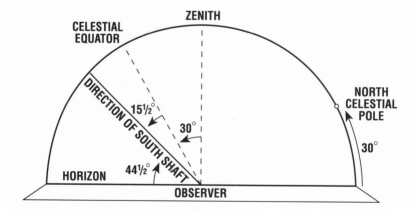

FIGURE 3. *Direction of south shaft in sky relative to positions of pole, equator, and zenith.*

It can be shown by spherical trigonometry that, for a star at declination δ and right ascension α (angular distance from the vernal equinox measured eastward along the celestial equator), precession will cause a change in position such that the declination at another time is given by

$$\sin \delta' = \cos \delta \cos a \sin \theta + \sin \delta \cos \theta$$

where $a = \alpha + \xi$, and θ and ξ are determined by the distance the ecliptic pole has moved due to planetary precession and the distance the north celestial pole has moved due to lunisolar precession during the given time. The values of these angles can be determined from the known rates and directions of the poles' motions. They have been tabulated for hundred-year intervals from 4000 B.C. to A.D. 3000 (for equinox 1900) by Paul Neugebauer[8] who has also worked out the right ascensions and declinations for 310 bright stars at hundred-year intervals from 4000 B.C. to A.D. 1900.[9] His tables and recent calculation by the same method show that one of the three stars in Orion's belt had a declination within 30' of $-15.5°$ (2840 to 2480 B.C.). The positions of the stars during this period were

Date	δ Orionis Declination	ε Orionis Declination	ζ Orionis Declination
3000 B.C.	−16°51′		
2900	−16°20′	−16°47′	
2800	−15°49′	−16°17′	−16°33′
2700	−15°17′	−15°46′	−16°05′
2600	−14°45′	−15°16′	−15°33′
2500	−14°17′	−14°6′	−15°04′
2400		−14°16′	−14°34′
2300			−14°6′

This means that these three stars, whose importance to the Egyptians we have seen, passed once each day, at culmination, directly over the southern shaft of the Great Pyramid at the time it was built.[*]

Thus considerations of Egyptian religion and modern astronomy combine to indicate that the "air shafts" of Cheops' Pyramid were actually in-

---

[*]This culmination was, of course, rendered invisible by daylight during about half the year. It would have been visible about 2700 B.C. from late July to early January.

tended as ways by which the soul of the deceased king might ascend to join the circumpolar stars and the god-constellation $\acute{S}\check{3}h$.

It would seem likely that some other stars might pass in the same fashion over the opening of the shaft. It happens, however, that no other stars of comparable magnitude had declinations within $1°30'$ of $-15°30'$ during that period.

Star of Bethlehem

The difficulties of adding apples to oranges are proverbial. But they are trivial compared to the difficulties of adding evidence from planetary orbits to evidence from the Books of Matthew and Luke. It is this latter task that astronomer David Hughes has essayed in attempting to identify the Star of Bethlehem with a particular astronomical phenomenon, a triplet of conjunctions of Jupiter and Saturn.[1]

Is such an attempt a sensible task? Hughes must have worried about this at great length, but he discusses it only briefly. The rejected alternatives to a normal astronomical event are (a) pure legend (midrash, Hughes calls it, using the Hebrew word in a relatively unfamiliar sense) and (b) true miracle, meaning an event neither predictable nor explicable in terms of the laws of physics.

Miracles and Midrashim

Hughes recognizes that births of heroic figures (including Abraham, Mithradates, and Alexander) and other important events in Near Eastern traditions are frequently supposed to have been foretold by the appearance of unusual stars. We should be careful, though, not to dismiss all prophetic stars as legendary just because they are common. Finegan has shown that there really was a bright comet (recorded by the Chinese) the year Mithradates was born (134 B.C.E.).[2] Hughes rejects the legendary alternative in this case on the grounds that the description of the star (Matthew 2:1–2) forms part of a narrative which has "the ring of truth," but is not sufficiently "coherent and plausible" to be invented. This narrative includes the Magi, the warning to Joseph in a dream of Herod's wrath, the flight into Egypt, and the Massacre of the Innocents. It does not include the shepherds, the stable and manger, or the multitude of the heavenly host, which appear only in Luke. Mark, generally regarded as the earliest synoptic gospel and as one of several sources available in common to the authors of Matthew and Luke, has no account of the birth or infancy of Jesus.

Hughes likewise rejects the miraculous alternative as "theologically weak because it requires a miracle when none was needed." This is apparently the statement of someone who believes that there are other, genuine, miracles recorded in the Testaments. Some of these appear gratuitous enough to merely human understanding; but we must suppose them to be necessary in the wider-seeing eyes of the Almighty, and I do not quite see how one can conclude otherwise about the star. Those of us who reject miracles in general must, *a fortiori,* reject a miraculous explanation of the Star of Bethlehem.

All three alternatives have been warmly espoused by reputable theologians in this century. Spend an afternoon poking among the BL (myth), BS (bible), and BT (theology) sections of any large library and take a vote of the authorities. Turn-of-the-century Germans are sure that the star is that of Numbers 24:17, bearing the same mythic interpretation as Bar Kochba's ("Son of a Star") name change on assuming leadership of the Jews; while Frazer supposes that Venus is meant; and McKenzie (complete with *nihil obstat*) opines that "the star is evidently described as wonderful and lies beyond any astronomical investigation."[3] For Van Dyke's Other Wise Man, we have more precise information. His journey was inspired by a double planetary conjunction with a nova in the middle. But then he never got to Bethlehem.

If the star is indeed to be interpreted as a natural astronomical phenomenon, then its relationship to the birth of Jesus is still not entirely clear. It might have been a real star incorporated into an otherwise legendary narrative for the sake of artistic verisimilitude. It could have been a natural phenomenon occurring at the same time as a miraculous birth through the action of divine providence. Or it could have been an astronomical event appearing by coincidence at the same time as a natural birth, the mythic connection being drawn only later by writers who had heard of both events secondhand. Hughes does not discuss these alternatives. He appears, however, to have one of the latter two in mind, as part of the purpose of his investigation is to pin down as closely as possible the date of Jesus' birth. It is, at any rate, not necessary to choose among these alternatives in order to follow Hughes' main arguments.

Let us then suppose with Hughes that the Star of Bethlehem is to be identified as some transient astronomical happening, near in time and space to the birth of an historical Jesus. How can we go about identifying and dating the particular event involved? Hughes adopts a multipronged approach that (a) derives the most likely properties of the star from the Matthew description, (b) narrows down the most likely date using data from both Matthew and Luke, (c) deduces what sort of phenomenon would most likely have interested Babylonian or Persian astrologers (if the wise men are so interpreted), and then (d) compares these criteria with the range of astronomical phenomena that are separately recorded or can be calculated to have occurred at the right time. There are difficulties at all four stages, which deserve to be looked at in detail.

The Star

Now, about that star. To begin with, it was apparently not very bright, for Matthew implies that Herod and his court had not been aware of it before the arrival of the Magi. The later writers of the epistle to the Ephesians (19) and the Protoevangelium (21:2)—and most designers of Christmas cards—claim astonishing brightness for the star; but Hughes regards the earliest description as most reliable.

The star either had a fairly long lifetime or appeared more than once, since it apparently motivated the wise men's journey to Jerusalem "from the East," and they rejoiced to see it (but whether "again" or "still" Matthew doesn't say) as they left Jerusalem for Bethlehem. The star was continuously visible during the latter journey, which could have taken at most a few hours. But Hughes believes that the star had disappeared during the weeks or months of the journey "from the East" (remember that the land immediately east of Israel was desert even then) and the Magi rejoiced because the star appeared *again* as they left Jerusalem, reassuring them that they were on the right track (like the signs for the George Washington Bridge that start appearing about the middle of Pennsylvania). It is unfortunate that classical Greek (like English but unlike some Amerindian languages) does not have separate verb forms for continuous and repeated actions.

A still longer life for the star may be implied by the Massacre of the Innocents. Herod asks the Magi when the star first appeared. Annoyingly, their answer is not recorded, but on the basis of it, Herod shortly thereafter (but we do not know how long) ordered the death of all children under the age of 2. Hughes takes this to mean that the star may have appeared about 13 months before Jesus' birth. The alternatives are that Herod was either a bit slow off the mark getting the massacre organized or very cautious. At any rate, a reasonably long life span (for the star, not the children) or a reappearance seems to be implied.

Where was the star in the sky? This used to worry me a good deal in childhood, for the King James version quotes the Magi as saying, "We have seen his star in the east." And if they were led to Jerusalem by a star seen in the east, they must surely have come from west of the city—which is the Mediterranean, and boats rather than camels would seem called for. I never solved that problem for myself, but Hughes offers two possibilities. First, the Magi may have meant "we have seen his star when we were in the east." This is plausible, but not very informative. And Hughes thinks it is wrong, on the grounds that a Greek writer would normally have used the plural, *anatolai*, to express "the east" as a location, rather than the singular, *anatole*, used by Matthew. The phrase, *en te anatole*, has specific astronomical significance, meaning "in the eastern sky," according to Clark *et al.*,[4] or "rising acronycally" (i.e., just as the sun was setting), according to Hughes. Notice that this doesn't assuage my childish puzzlement that the Magi having seen something to the east of them should promptly start out marching to the west, but it will turn out at least to be an interesting constraint on the identification of the object. By way of further constraint, Hughes believes that when Herod asked the wise men "what time the star appeared" he meant "on what date did it rise heliacally" (just ahead of the rising sun), so that the relevant astronomical phenomenon must have been capable of that as well. Notice that if the star had a fixed position, this would have put the Magi in the awkward position of wanting to answer Herod with "six months earlier, you twerp; can't you count" while realizing that some more politic response was in order.

Finally, and most fascinating of all, during the Magi's journey from Jerusalem south to Bethlehem, "the star, which they saw in the east, went before them, till it came and stood over where the young child was." The behavior is not literally within the capabilities of any astronomical object. Hughes, looking forward to his eventual choice of events, suggests that the most likely meaning is that the star was on the southern meridian during the journey from Jerusalem to Bethlehem ("going before") and quite close to the zenith at that latitude ("standing over"), with a little midrashic embroidery on the part of the narrator. Hughes' interpretation is at least a physically possible description that doesn't violate the other constraints: fireballs or aurorae might more plausibly "go before" or "stand over," but they do not last long enough to fulfill the duration conditions. But Hughes is, in effect, making a decision to take some portions of the narrative literally and others not. He never presents a general prescription for how to make such a decision, and I don't see how it can be done (in this case or others that will arise) except in the light of a particular hypothesis about what the star was. To say that the hypothesis is consistent with all the data then becomes somewhat circular.

Thus, in summary, the star (a) was not very conspicuous, (b) lasted at least six months or reappeared or both, (c) could be seen at the eastern and western horizons and at the zenith (not all at the same time!), and, perhaps, (d) moved around in unstarlike ways. At no point in the narrative is the star referred to except by the singular of the ordinary Greek work for star *(aster),* which differs both from the plural and from the word for planet (wanderer). Matthew appears to have thought he was talking about a single, non-wandering star.

The Date

When did all this happen? The writers of Matthew and Luke both attempted to provide "time" frames for their narratives. Unfortunately, not everything they say can be simultaneously and literally true. The second chapter of Luke begins "And it came to pass in those days [meaning, presumably, the days of chapter 1, which deals with the conception and birth of John the Baptist, when Herod was King of Judea] that there went out a decree from Caesar Augustus, that all the world should be taxed." That's easy. Augustus was more or less in charge of things from 36 B.C.E. to 14 C.E., covering all likely dates for both star and birth. No independent record of his having taxed Syria and Judea or having taken a census for the purpose of taxation has so far come to light. But we do know that the Jews did, in fact, pay some Roman taxes from 63 B.C.E. onwards. Hughes reports an inscription from Ankara listing irregularly spaced years in which tax collections were ordered there. The only likely date is 8 B.C.E. In addition, Acts 5:37 reports another census for taxation purposes in the 37th year after Caesar's victory over Anthony at Actium (6 C.E.). This, in combination with records of censuses carried out at 14-year intervals in Roman Egypt (34, 48, 62 C.E.), also points to a decree in 8 B.C.E.

So far, so good. But Luke 2:2 continues, "And this taxing was first made when Cyrenius was governor of Syria." Now existing Roman records put P. Sulpicius Quirinius in the governor's chair in 6–7 C.E. and perhaps also in 3–2 B.C.E. Neither is a good match to 8 B.C.E., and both are too late to qualify as the days of King Herod. Hughes explores a couple of ways out of this contradiction. We can "correct" Luke to read "This taxing was the first before that made when Cyrenius was governor," or some other decoupling events. Or we can suppose that Luke's information was slightly faulty and the census and taxation took place under Saturnius (governor 9–6 B.C.E.) or Varus (6–4 B.C.E.), with Quirinius assisting. Tertullian, writing in the second century, spoke of "the census taken in Judea under Augustus by Sentius Saturnius," but his having known which governors overlapped with Herod is not really evidence for anything. The basic contradiction is that neither of Cyrenius' spells as governor occurred in the days of Herod, King of Judea. And we cannot save the situation by supposing Luke to have meant the next King Herod, Antipas (Claudius' friend, to Robert Graves fans), for he reigned until 38 C.E., far too late to bring the holy family back from Egypt with his death. Thus 8 B.C.E. is a perfectly plausible census date, which, allowing for slowness of transport and communication, would presumably bring Mary and Joseph to Bethlehem in 7 or 6 B.C.E., but neither it nor any other date for the taxation decree can be adopted without modifying or ignoring at least one of Luke's statements.

Matthew complicates the situation further by telling us (a) that Jesus was born in the days of Herod, (b) that Herod decreed (how much later is unclear) a slaughter of infants, prompting the flight into Egypt, and (c) that Herod then died, allowing the return from Egypt. A firm date for Herod's death would thus provide a *terminus ad quem* for Jesus' birth. Although historians writing at Rome took little interest in provincial events and do not mention the death of Herod, his successors to the various pieces of the dismembered Judean kingdom do appear in Roman documents and must have assumed their offices in 5–4 B.C.E.[5] Josephus, the one contemporary historian to mention the death, records a lunar eclipse visible at Jerusalem, followed by various Herodian misdeeds and illnesses ending in death, followed by Passover. Josephus was alive and well at the fall of Massada (76 C.E.), thus not quite contemporary, and a Jew who went over to the Roman side just in the nick of time, thus perhaps not quite unprejudiced. The sins of historians so circumstanced are notorious, as for instance Thomas More writing on Richard III (see Josephine Tey's *The Daughter of Time* for an exposé of the midrashim in that history). And there are more recent examples. But dates of eclipses and Passovers can be calculated. The eclipses of 15 September 5 B.C.E., 13 March 4 B.C.E. (partial), and 10 January 1 B.C.E. were visible from Jerusalem; the subsequent Passovers fell on 11 April 4 B.C.E. and 8 April 1 B.C.E. Most writers have chosen the middle eclipse, killing off Herod in

March or April 4 B.C.E. Hughes opts for the earlier eclipse, so as to allow more time for the events recorded by Josephus and for the comings and goings of the Magi and the holy family as reported by Matthew. He terminates Herod on 5 or 6 December 5 B.C.E. The date of Jesus' birth is thus constrained to 8–5 B.C.E.

Hughes further discusses the time required for the Magi's journeys, the age of Jesus when they arrived, the events surrounding the birth of John the Baptist, and other issues, eventually choosing 7 B.C.E. as the most likely birth year. Again, this is perfectly plausible, but it is arrived at by taking some of the New Testament narrative more seriously than other parts which seem to me equally credible. Finally, the second appearance of the star, leading the Magi from Jerusalem to Bethlehem, also ends up in 7 B.C.E., while the first appearance "in the east" comes earlier by four months (journey time) to two years (age of infants slaughtered).

The Magi

The Magi (Greek *magoi*) can properly be called wise men (King James version), astrologers, or magicians in English. That they were kings and three in number is not implied by the Matthew narrative, and their traditional names and ethnic backgrounds appear only in the fifth to seventh centuries. Hughes describes assorted magi known to Old Testament, Greek, Roman, and later authors. He concludes that the visitors to Bethlehem were most probably astrologers from Babylonia, where Hebrew prophecies would be well known because of the large Jewish population remaining there from the first exile. Persia, Parthia, and southern Turkey (Commagene) are less likely starting points. The location hardly matters to the chronology: a journey from any one of these places could have taken anything from two weeks (by camelback, according to T. E. Lawrence) to a year (extrapolating from Ezra's return to Jerusalem).

But if the Magi were Babylonian astronomers/astrologers, they would have been much more likely to be interested in certain kinds of phenomena than others. It is at this point that Hughes diverges most strongly from some other astronomers who have recently written about the star.[6] They have supposed that the Magi would most likely have taken notice only of an unexpected event, like a comet or a nova, while Hughes believes that only predictable events, like planetary configurations, would have had sufficient astrological significance for the Magi to deduce the birth of the king of the Jews. Deciphered cuneiform inscriptions show fairly definitely that the first-century Babylonians could accurately predict planetary positions (including perhaps even the conjunction that Hughes eventually identifies as the star). I do not know how to choose between the alternatives of surprising and predictable events.

Our modern scientific prejudice is that "predictable" very nearly equals "understood" and so "uninteresting." But our own prejudices are very poor guides to the feelings of even our parents, children, or overseas contemporaries, and must be nearly useless for entering into the mentality of Babylonians 2000 years ago. So take your choice—but carefully, as it plays major role in the final identification.

Astronomical Events

The realm of the fixed stars is full of things that change. Transient (or transiently visible) astronomical phenomena suggested in the past for the star of Bethlehem include aurorae, ball lightning, meteors and meteor showers, zodiacal light, Venus, variable stars (especially Mira), and Canopus as first glimpsed by someone going south from northern Assyria. Hughes gives all of these fairly short shrift: some because they do not last long enough; others because they cannot "go before" and "stand over" (but this applies also to the winning choice); still others because they would have been too conspicuous for Herod not to have known about them before the Magi arrived; and all because they "lack positive astrological significance," and so could not have heralded the prophesied birth in the minds of the wise men.

Comets, novae, and supernovae, recorded by the Chinese as "guest stars" and "broom stars," merit more serious consideration, the more so as standard compilations record guest stars, apparently of long duration, in the springs of 5 and 4 B.C.E. Clark *et al.*,[7] Morehouse,[8] and others, supposing surprise to be of the essence for the star, have presented strong cases for one or the other or both of these. But Hughes in the end rejects them too. The guest stars of 5 and 4 B.C.E. were too conspicuous for Herod to have missed (but remember the curious incident of the Europeans in 1054!), occurred too close to his death, and most importantly, again lacked astrological significance for the Magi.

Only one hypothesis survives, that of an unusual planetary configuration. This will meet the criteria of relative inconspicuousness, duration, and astrological significance, though not those of literally standing over, going before, or being best described as a single *aster*. Modern calculations show that a dozen or so planetary conjunctions (two planets at the same longitude) and massings (three or more planets close together in the sky) occurred during the reign of Augustus. Only one of these falls in the critical 5–4 B.C.E. period, a triple conjunction of Jupiter and Saturn (meaning that the planets cross in longitude three times, owing to a combination of their motion and the earth's) in May, September, and December 7 B.C.E., while the planets were in Pisces. Such triple conjunctions occur only every 139 years, and ones in Pisces (which constellation, Hughes notes, was particularly associated in astrological thought with Israel and the Last Days) only about every 900 years—a very rare event, but a predictable one. The 7 B.C.E. conjunction was known to Keppler, who identified it and the associated massing of Jupiter, Saturn, and Mars (February 6 B.C.E.) with the Star of Bethlehem. It has been discussed as such on and off ever since.

The planetary hypothesis is not without difficulties of its own. The 7 B.C.E. conjunction was not a particularly close one. The planets remained at least 1° apart throughout and can never have looked like a single *aster*. In addition, learned Greeks of the period were well aware of the difference between stars and planets and had a perfectly good word for the latter (*planes*), which Matthew does not use.

The triple conjunction cannot literally have led the Magi or pointed out a single house. It matches the author's interpretation of "going before" and "standing over" only rather approximately. The planets, of course, crossed the southern meridian at Jerusalem daily, close to dawn in May, near midnight in September, and in early evening in December. But they can never have been at the zenith at Bethlehem, which has a latitude of about 30°, while the planets are confined to the ecliptic. Thus the December meridian crossing would have been about 8° from the zenith and the others further away.

Finally, the conjunction cannot really be said to have disappeared and returned to cause the Magi great rejoicing as they left Jerusalem. Jupiter and Saturn were visible at some time of night continuously from heliacal rising in March 7 B.C.E. to heliacal setting in March 6 B.C.E. On the other hand, the planets did move to about 3° apart in the summer and approach again to within about 1° of each other in the fall. Perhaps that is the second appearance of the star. Hughes worries about all of these issues, but in the end decides to give greater weight to astrological significance than to the details of Matthew's description.

Deductions, Alternatives, and Conclusions

If Hughes is right in associating the 7 B.C.E. triple conjunction with the Star of Bethlehem, then the birthday as well as the birth year of Jesus is somewhat constrained. Hughes examines possible associations of conception date and birth date with the significant aspects of the conjugation—heliacal rising, acronycal rising, and minimum separation of the planets. He opts tentatively for birth on the day of acronycal rising, 15 September 7 B.C.E., the Magi arriving sometime in late November or December. This choice corresponds to the acronycal interpretation of the passage "We have seen his star in the east." Other combinations are clearly possible, and Hughes does not seem to feel very strongly about the issue. One might, for instance, associate the March heliacal rising of the planets with the conception, the Magi arriving just in time for the birth nine months later, guided to Bethlehem by the third conjunction on the evening of, say, 24 December. It is only fair to admit, though, that the 25 December tradition does not seem to go back earlier than the fourth century.

Has Hughes really exhausted all the possibilities? It might be interesting to cast the net wider and look at astronomical events throughout the reign of Augustus. Luke and Matthew almost certainly knew who was emperor, but, like a good many of us, may have been a bit vague about who their governors and senators were at any given time.

Some of the astronomical phenomena may have been dismissed too quickly. That cases can be made for assorted "guest stars" has already been noted. Many of us have a Giotto-and-Christmas-card prejudice in favor of a comet with its tail pointing right down to the stable. Admittedly, the geometry of this is more convincing in two dimensions than in three, but who is to say that the collective artistic consciousness of mankind may not preserve a truth that has otherwise been lost. I have another, private, favorite interpretation. Whipple showed many years ago that the material that now gives rise to the Taurid meteor shower must have broken off from the comet Encke "about the year zero."[9] There is a year zero in astronomical terminology (though historians call it 1 B.C.E.); and the calculation involves the differential precession of the comet and meteor stream orbits. Ancient watchers of the sky must have known the recurrent meteor showers well and would have noticed a new one. There is nothing sensible to be said about the astrological significance of such an event, but if the Magi saw the new shower for the first time in Persia or Babylon and then saw it again a year later in Judea, they would surely have thought of it as the same object, because the meteors would be coming from the same place in the sky (radiant). Calling a few thousand shooting stars *aster* may be stretching a point, but this sort of event does about as well (or as badly) as any other in matching the details of the Matthew description.

How much of all this ought we to believe? It seems beyond reasonable doubt (a) that there was a triple conjunction of Jupiter and Saturn in 7 B.C.E. and (b) that the astronomers/astrologers of the period were accustomed to observing and predicting such events with some precision. That there must also have been naked-eye comets and/or novae during the reign of Herod we would expect on statistical grounds, even without the Chinese records. The same holds true for meteors and the rest.

But the real questions are harder to answer: Did the author of Matthew actually have in mind a specific, natural astronomical event? And if so, did it in fact bear the same temporal (and causal) relationship to the birth of a historical Jesus that he thought it did? If the answers to these are "yes," then Hughes has presented a rather convincing case that the particular event was the 7 B.C.E. triple conjunction. But neither he nor any other astronomer can really answer the two harder questions for anyone except himself; nor does it seem likely that future archaeological discoveries or textual criticism will shed more light on the problem. You must (as Nero Wolfe used to say to Archie Goodwin) use your intelligence guided by experience. My guess is "no" to both; but then I'm Jewish.

Hughes has attempted an extraordinarily difficult task, both intellectually and philosophically. I cannot imagine that anyone else could have done it better. But in combining the apples of New Testament scholarship with the oranges of planetary astronomy (and a few astrological sour grapes), he has produced a sort of fruit salad, which, like most salads, should probably be consumed only after liberal application of sodium chloride.

In Their Own Words

Nearly all of the major ideas of modern cosmology belong to our own century, following upon Hubble's 1925 recognition of the expansion of the universe. Important components include dark matter, gravitational lensing, very large scale structure (scale, amplitude, topology, origin), galaxy formation, and the microwave background and cosmological nucleosynthesis. Many of the pioneers of these ideas are still actively engaged in astronomy; others left us only recently. As a result, history of cosmology often falls within the realm of the after-dinner speaker, whose memory may be faulty in two ways. First, he may forget some of the things that happened, and second, (in the words of Freeman Dyson) he may remember ones that did not.

Thus I set out to collect short verbatim extracts from early papers, most of them by very well-known astronomers, for comparison, if the reader wishes, with the versions he remembers or remembers hearing after dinner. Each section stops roughly when someone has published what we now regard as the right answer.

The starting points are determined by a friends-of-friends algorithm. That is, all the authors directly quoted are people I have met either first or second hand. This is not so restrictive as it sounds. If you have had the pleasure of meeting Jan Oort, then you know Kapteyn and de Sitter at one remove. Similarly, William McCrea "introduces" you to Milne and Jeans, Peter Bergmann to Einstein, and so forth. The purpose of this restriction is to stay close enough to our own times and paradigms for words to carry essentially their current meanings. My own linguistic limitations force a focus on papers in English or English translation.

Though only formal publications are quoted here, my mental map of the territory derives heavily from conversations and after dinner talks. Important contributors have been George O. Abell, Horace W. Babcock, George Gamow, James E. Gunn, Vera C. Rubin, Joseph Weber, Olin C. Wilson, Yakov B. Zel'dovich, and Fritz Zwicky. Where possible, the references give the page on which the quoted remark occurs, not the first page of the paper.

Dark Matter

Velocity data over a wide range of length scales have now persuaded most astronomers that dark matter (stuff that contributes to gravitational potential without radiating its fair share of light) dominates the dynamics of galaxies, clusters, and the universe, without telling us what it is made of. A discussion of such dark matter formed part of Kapteyn's first attempt, in 1922, at a theory of the structure and dynamics of the galaxy. He derived mass in the galactic plane from the distribution of star numbers and velocities perpendicular to it and concluded: "We have the means of estimating the mass of dark matter in the universe. As matters stand at present it appears at once that this mass cannot be excessive. If it were otherwise, the average mass [per star] as derived from binary stars would have been very much lower than what has been found for the effective mass."[1] In the same year and using much the same data, Jeans decided that "there must be about three dark stars in the universe to every bright star."[2] Their "universe" was, of course, our "galaxy," as Oort knew a decade later when, with improved data and theory on "the force exerted by the stellar system in the direction perpendicular to the galactic plane" he found that "one unit of light corresponds to 1.8 solar masses."[3] These determinations bracket the modern values.

On the scale of individual galaxies, Babcock (1939) was first to follow a rotation curve (M31) far enough out to see "the ratio of mass to luminosity, in solar units, is about 50"[4] (17 for the modern distance scale). He has said that the reception of his conclusions, both at his graduate institution and at an AAS meeting, was such as to persuade him to a career well away from extragalactic astronomy.[5] His published thesis makes no mention of the two following results, which also indicate large masses for galaxies.

Holmberg (1937) pioneered in taking binary galaxies seriously as bound systems and said, "If the orbits are assumed to be circular we now arrive at the following value: Average mass of the galaxies = $1.0 \cdot 10^{11} \odot$"[6] (which is 5×10^{11} $M_\odot$ for H = 100 km/s/Mpc).

On still larger scales, Zwicky (1933) collected enough red shifts for galaxies in the Coma cluster to estimate M/L = 400 (old distance scale) and to opine "dass dunkle Materie in sehr viel grösserer Dichte vorhanden ist als leuchtende Materie" ("that dark material is present in much larger quantities than luminous material").[7] Sinclair Smith, using a spectrograph of his own design and construction, brought the total number of Virgo red shifts up to 32 and found that "the mass per nebula is $2 \cdot 10^{11} \odot$"[8] (which is $10^{12} M_{\odot}$ for H = 100). Notice that when all are expressed in a consistent distance scale, the binary galaxy value nestles comfortably between the M31 and the cluster numbers.

A sort of renaissance of dark matter occurred at a 1961 conference in Santa Barbara, where Abell presented "evidence for the existence of second-order clusters" and said, "Typical masses for these systems are estimated at from 10^{16} to 10^{17} solar masses."[9] At this same meeting, Von Hoerner presented what may have been the first n-body simulation of a cluster. He concluded that "deviations from the time averages in the virial theorem could be no larger than this [a factor of two]."[10] His n was 16! Alternatives to dark matter discussed at the conference included unbound expanding clusters (advocated by Ambartsumyan) and substructure, interlopers, and observational errors (Holmberg). The first suggestion of an alternative in the form of "a law of gravitation that implies a much stronger attraction at the long distance than that predicted by the law of Newton" came two years later.[11]

The end of the folklore period in this subject came in 1974 with a pair of papers that summarized the existing data as "the mass of spiral galaxies increases almost linearly with radius to nearly 1 Mpc"[12] and "the mass–luminosity ratio rises to $l \approx 100$, $l \gtrsim 120$ for elliptical galaxies."[13] In passing, we note (a) an early cosmological limit to the mass of the muon neutrino of 400 eV/c^2 [14] and (b) the coining of the name WIMP.[15]

Gravitational Lenses

Gravitational lensing would presumably result from Newtonian gravitation bending the paths of Newtonian light particles; but the first discussions of it as an astronomical phenomenon in fact came in the wake of general relativity and the detection in 1919 of the deflection of starlight by the sun. I have found some difficulty in disentangling the early history and so begin in the middle. Zwicky pointed out in 1937 "that extragalactic nebulae offer a much better chance than stars for the observation of gravitational lens effects"[16] and that "observations on the deflection of light around nebulae may provide the most direct determination of nebular masses."[17] He was right on both counts, in that the only lensing we have unambiguously seen is that of one galaxy (or quasar) by another galaxy or cluster and that often the implied lens mass is quite high. And it took only about 50 years from promise to fulfillment!

Zwicky cites the (fairly) well-known 1936 paper by Einstein (it addresses lensing of and by stars only), but Zwicky says that he had the idea of "image formation through the action of gravitational fields" from V. K. Zworykin, to whom it had been suggested by R. W. Mandl.[18] Einstein's attribution is "Some time ago, R.W. Mandl paid me a visit and asked me to publish the results of a little calculation which I had made at his request."[19] But there are two confounding earlier items. First, the earliest lens paper I have found ("Über eine mögliche Form fiktiver Doppelstern," or "Concerning a possible form of fake double star," by Chwolson, 1924), fills only about a quarter of a page and is followed immediately by an even shorter note on the index of refraction of an electron gas—by A. Einstein.[20] And it is hard to believe that he didn't at least glance at that page at some time and so see the Chwolson note. Second, Zwicky footnotes that "Dr. G. Strömberg of the Mt. Wilson Observatory kindly informs me that the idea of stars as gravitational lenses is really an old one. Among others, E. B. Frost, late director of the Yerkes Observatory, as early as 1923 outlined a program for the search of such lens effects among stars."[21] Only on a late rereading did I finally turn over the page with Chwolson and Einstein, thinking that surely the last short paper would be of no interest. But it bears the name Leigh Page, father of our colleague Thornton Leigh Page. (And the grandson is Leigh Page II.) The world is a small and sometimes confusing place!

Very Large Scale Structure and Streaming

Recognition that the spiral nebulae are concentrated in a supergalactic plane on the sky predates understanding of their nature and is usually credited to the younger Herschel (not a friend of a friend). The hypothesis of local mediocrity having become fashionable only rather recently, it seems fair to date the discovery of superclusters to 1934 when Shapley found the second one, saying, "If it were not for its great distance. . . the twin supergalaxy in Hercules . . . would be equally interesting [to Virgo]. Future studies of this region of the sky may show that in fainter members the twins are one, and show also that they are but concentrations in an exceedingly rich region, a metagalactic cloud."[22]

The existence of superclustering long remained in dispute, particularly by Zwicky, who in 1937 reported "that practically all nebulae must be thought of as being grouped in clusters"[23] but maintained to the end that "the concept of secondary clustering, in our opinion, thus may be discarded as superfluous."[24] And I remember George Abell in the mid-1960s offering Herzog "political asylum" at UCLA when he had presumed to find such structure in the Zwicky, Herzog, and Wild *Catalogues*. But in fairness one should say that Zwicky's unit of inhomogeneity was the "cluster cell" whose "indicative diameter . . . based on a red shift constant of 100 km/sec per million parsecs is . . . 41 million parsecs,"[25] more or less the size we now attribute to superclusters, as Zwicky and Rudnicki themselves noted.

The amplitude and range of large scale deviations from uniform Hubble flow have remained controversial to our own time. Whatever the eventual outcome, both credit and gratitude in large quantities belong to those who first took the issue seriously. Among the foremost is Vera Rubin. Cognoscenti will think first of the Rubin–Ford effect.[26] But still more prescient is Rubin's 1951 paper which reported a peculiar motion of our galaxy of 179 km/s toward $l = 105°$, $b = +39°$ relative to galaxies with velocities up to 1000 km/s and "a peculiar motion of our galaxy of 373 km/sec toward $l = 160°$, $b = +22°$" relative to galaxies with velocities from 1000 to 4000 km/s.[27]

Finally, one cannot close this section without mentioning Gerard de Vaucouleurs, an early and firm proponent of superclustering and also of what we would now call Virgocentric infall, in the form of "a differential expansion effect explainable by the hypothesis that the expansion rate is small or negligible in the denser central regions and increases asymptotically outwards in regions of decreasing density . . . around a center located in the Virgo cluster."[28]

Galaxy Formation

Since we do not know the answer to how galaxies formed (or rather many people know the answer, but they know different ones) this section has no coherent focus, though I think the line between "history" and "current events" can be drawn fairly cleanly in 1962, when Eggen,

Lynden-Bell, and Sandage published "Evidence from the Motions of Old Stars that the Galaxy Collapsed."[29]

The basic idea goes back at least to Jeans, in 1929, who said (crediting Kant and Laplace) that "the picture to which we are naturally led is that of matter scattered uniformly through space. It is satisfactory to find that if this matter were in the gaseous state, its molecules moving with reasonable velocities of thermal agitation, then the next stage in the evolution of the universe would be the formation of distinct aggregations having masses comparable with those of the spiral nebulae" via what he called the gravitational instability and we now call the Jeans instability.[30] That this still occurs in an expanding universe is not self-evident but was demonstrated by Lifshitz[31] and Bonnor,[32] who went on to say that "the condensation process is too slow to account for the formation of the nebulae," which is true if you start with only statistical fluctuations.

Von Weizsäcker discussed formation of stars (with planets) and galaxies together[33] and said that "galaxies seem to have been formed by a competition between expansion and turbulence" and that "irregular nebulae must be young, spirals intermediate, elliptical galaxies genetically old" (though many of his contemporaries envisioned ellipticals gradually spinning up to spirals as revealed in the terminology "early" and "late" galaxies). Hoyle's 1953 concept, now called hierarchical fragmentation, is closely related and predicted "that galaxies are likely to form, not as single objects, but in clusters."[34]

Gamow returned several times in the 1940s and 1950s to the difficult problem of making galaxies in a hot, expanding universe, proposing first that "the type of expansion necessary for the formation of nebulae indicates that space is infinite and unlimitedly expanding,"[35] somewhat later that "shadow casting" or "mock gravity" can "effectively increase the attraction between various parts of the expanding gas and could be made responsible for the formation of rudimentary condensations,"[36] and that "the primordial gas uniformly filling the universe during the early stages of its expansion must have been broken up into separate giant gas clouds (protogalaxies) during the transition epoch . . . from the predominantly radiative into the predominantly material state . . . when the density of gas and the mass density of radiation were comparable."[37]

Bonnor being absolutely right that statistical fluctuations cannot grow into galaxies in the available time,[38] the origin of the primordial fluctuations has received much attention. Just before the discovery of the 3 K background radiation in the mid-1960s, Sakharov (whom we can all claim as an acquaintance once removed through M. M. Shapiro) suggested "that the initial inhomogeneities arise as a result of quantum fluctuations of cold baryon–lepton matter at densities of the order of 10^{98} baryons/cm^3,"[39] that is, at the Planck time. The first fluctuations to separate out in his model are the globular clusters, so it is an early "bottom-up" scenario.

The Microwave Background and Cosmological Nucleosynthesis

Alpher and Herman have published an account of their own participation in the development of this subject and should be regarded as the primary source.[40] The "prediscovery" observation of excited interstellar CN by W. S. Adams is widely known. McKellar reported in 1941 that it implied "for the region where the CN absorption takes place, the 'rotational' temperature T = 2.3 K."[41] He remarked that the value is quite insensitive to the exact line ratio, while Herzberg quotes the result as "a rotational temperature of 2.3 K . . . which has of course only a very restricted meaning."[42] (This was one of our favorite phrases in graduate school just after Penzias and Wilson's *Astrophysical Journal* letter[43] appeared.)

The famous Alpher, Bethe, Gamow paper really does exist. It occupies rather less than two pages in the April 1, 1948, issue of *Physical Review*. Nucleosynthesis is described as successive neutron captures, beginning with deuterium, "just above the upper fringe of the stable elements" followed by beta decays, so that "individual abundances of various nuclear species must depend not so much on their intrinsic stabilities (mass defects) as on the values of their neutron capture cross sections."[44] No mention is made of pervasive radiation or of a present temperature. This first appears in the November 13, 1948, issue of *Nature*, where Alpher and Herman report, "The temperature in the universe at the present time is found to be about 5°K."[45] But this is still based on a universe consisting initially of pure neutrons and building all the elements by successive captures thereof. Details of the calculation followed shortly.[46] It is arguably significant that, in 1949, Gamow assured an aspiring graduate student with a background in microwave spectroscopy that he knew of no interesting thesis problems in that field.

Hayashi recognized in 1950 that at sufficiently high temperature the *n/p* ratio was determined by rates of electron and positron captures and emissions. He found that the ratio should be *n/p* = ¼ at the beginning of nucleosynthesis and that "the hydrogen–helium abundance ratio (in numbers) resulting from the initial n-p ratio 1:4 become 6:1" ($Y = 0.4$). He also hoped that the presence of free protons would help to bridge the $A = 5$ and 8 gaps.[47]

When American and Soviet theorists began thinking seriously about the early universe again in the 1960s, it was in the wake of a paper by Ohm that reported a 2390 MHz atmospheric temperature very close to what was expected theoretically (I think it came from extrapolating $csc\ \theta$ dependence to $csc\ \theta = 0$ and so was bound to). This was widely misinterpreted. But a couple of pages later in the article comes a discussion of the contribution of the horn antenna to the system temperature, 2 ± 1 K. And "this estimate is based on the temperature 'not otherwise accounted for' in a previous experiment; it is somewhat larger than the calculated temperature expected from back lobes measured on a similar antenna."[48] This is evidently where the background is hiding.

Taking the Ohm result to imply an upper limit of 1 K to the cosmic temperature, Zel'dovich, for instance, concluded in the early 1960s that a hot big bang was ruled out because "such a density of radiation is obviously in contradiction with both radio-astronomical observations and the indirect data of cosmic-ray theory."[49] (The latter is still a bit of a problem.) He turned his attention (though not that of all his colleagues) to a cold big bang, consisting initially of only protons, electrons, and neutrinos. Doroshkevich and Novikov had just published a prediction of the "spectrum of the metagalaxy" from 10^9 to 10^{15} Hz,[50] and Dicke and his colleagues were about to do so in 1964, when Penzias and Wilson collected the data they reported as "A measurement of Excess Antenna Temperature at 4080 Mc/s,"[51] bringing this subject, and, really, all of cosmology, into the modern era.

Relativistic Astrophysics

Introduction and Disclaimer

Relativistic astrophysics began in 1963. The rather unusual precision of the starting point arises from the name and discipline having both been deliberately invented in connection with a conference convened to discuss interaction between observations and theories of quasars and related objects. In February 1963, Fred Hoyle and William Fowler suggested that the energy source for powerful extragalactic radio sources might be the gravitational collapse of superstars with masses 10^5 to 10^8 that of our sun.[1] And in March 1963, Maarten Schmidt and others determined redshifts for the optical counterparts of two very strong, compact examples of these sources.[2] The redshifts placed these two quasi-stellar radio sources (QSRSs, later quasars) at the outskirts of the known universe, implying that they must have both radio and optical luminosities hundreds of times larger than those of normal galaxies, coming primarily from a volume less than a light year across.

A handful of experts in general relativity quickly recognized the potential applicability of their specialty to these objects, and in June, Peter Bergmann, Ivor Robinson, Alfred Schild, and Engelbert Schucking invited a number of astronomers, relativists, and other physicists to participate in a symposium which they eventually entitled "Gravitational collapse and other topics in relativistic astrophysics." According to the published proceedings, about 300 scientists participated in what became retrospectively numbered as the first of the Texas Symposia on Relativistic Astrophysics (of whom 291 are listed by name in the proceedings, presumably because they formally registered).[3] This gathering actually took place in Texas (Dallas) on 16–18 December 1963, in the wake of the Kennedy assassination, and the series has continued (generally under happier circumstances) through the 15th Texas Symposium in December 1990 in Brighton, England.

History and demography of science and scientometrics are, of course, separate academic disciplines, with their own paradigms, graduate programs, literature, departments, and customs. I can claim no credentials or special expertise in any of them. My background is in observational astronomy and data interpretation, and my publications deal primarily with white dwarfs, supernovae, binary stars, dark matter, and other purely astronomical topics. This examination is, therefore, best described as part of the folklore of science—the collection of stories and memories we share with our students by way of initiating them into the community.

Such folklore is heavily subject to what I should like to call the Dyson effect, from a remark by Freeman Dyson (in a talk discussing his early association with Richard Feynman) that, when he looked back at letters to his parents describing the events of 40 years before, he was amazed both by how much he had forgotten and by how much he remembered that never actually happened. My own memory of the Texas Symposia I have attended (all but 1, 2, and 8) is surely at least as fallible. But published proceedings exist for Symposia numbers 1,[4] 2,[5] and 6–14 (numbers of the *Annals of the New York Academy of Sciences,* apart from the 13th Symposium[6]); my nearly verbatim notes (taken at the rate of about five pages per hour of talks) at symposia 3–7 and 9–15 have provided some additional contemporaneous information.

In retrospect, the discovery of extra-solar-system x-ray sources in 1962–63,[7] which prompted a very large number of calculations of neutron star cooling, and the rapid realization that the Crab Nebula source was not actually compact[8] were nearly as important to the crystallization of relativistic astrophysics as was the early work on quasars and their interpretation.

The Subject Matter of Relativistic Astrophysics

The initial focus on applying general relativity to quasars and extragalactic radio sources rapidly broadened to include both other astronomical phenomena and other portions of theoretical physics. At times, it has looked as if relativistic astrophysics might try to take over all of astronomy and physics. The sixth Texas symposium, for instance, included chemical fractionation of the interstellar medium ("What do you mean, relativistic molecules?!" one colleague exploded) and the eleventh, large telescope design (though Newtonian mechanics is more than adequate to describe the motions of even the largest ones now contemplated).

What has, in fact, happened is not quite this uncontrolled growth. New objects were added to quasars either upon discovery or when it came to seem likely that special or general relativity was required for their understanding. And new parts of physics have been incorporated when they seemed pertinent to supernovae, the early universe, or other items already in the inventory. A good deal of territory has also been abandoned over the years.

Prehistory

A double handful or so of the objects, processes, and concepts that we now think of as within the domain of relativistic astrophysics had been studied, or at least thought about, well before 1963. These include white dwarfs, neutron stars, black holes, and galaxies with spectacular nuclei; the processes of synchrotron radiation, accretion-powered x-rays, and gravitational lensing; and the ideas that events occurring very early in the history of the universe might have left trackable footprints on the present and that interpretation of the statistics of radio sources requires general relativistic calculations at a fairly sophisticated level. One might also in this context think of cosmic rays and early evidence for dark matter in galaxies and clusters.

Werner Israel has carefully summarized the discovery and understanding of white dwarfs, neutron stars, and black holes and provides extensive references.[9] A noteworthy point is that, while white dwarfs were observed both indirectly (F. W. Bessel, 1844) and directly (A. G. Clark, 1862) and recognized as puzzling (W. S. Adams, 1914) long before they were modeled (S. Chandrasekhar 1931–35), both black holes (J. Michell, 1784, and P. S. de Laplace, 1799) and neutron stars (W. Baade and F. Zwicky, 1934; J. R. Oppenheimer and G. Volkoff, 1939) were discussed as theoretical entities well before they were first seen—neutron stars as pulsars by S. J. Bell and A. Hewish in 1967–68, and the first persuasive black hole candidate as a component of the x-ray binary Cygnus X-1, with orbit parameters pinned down by C. T. Bolton in 1972.

The history of neutron stars provides an interesting example of the Dyson effect. Baade and Zwicky were first into print with the idea that supernovae and cosmic rays are powered by neutron star formation,[10] but the actual invention of the objects has generally been attributed to

Lev Landau, on the basis of Leon Rosenfeld's after-dinner-talk memory of a conversation between the two of them and Bohr, in Copenhagen in 1932, triggered by the receipt of a letter from England announcing Chadwick's discovery of the neutron. But Bohr's old institution keeps its files, and Gordon Baym, searching among them several years ago, found that neither Landau nor Bohr was actually in Copenhagen at the time of the discovery of the neutron. Landau had already returned to the Soviet Union, and Bohr was elsewhere in Europe. Israel suspects that the remembered conversation probably took place in Kiev in 1935—after Baade and Zwicky, who may have been the real inventors.[11]

Seyfert galaxies (a very mild form of quasar in which the compact nucleus is about as bright as the rest of the galaxy instead of 100 times it) are genuinely named for their discoverer.[12] By 1959, L. Woltjer had recognized (a) that their numbers imply lifetimes as Seyferts only 1 percent of the age of the universe and (b) that confinement of the rapidly moving gas responsible for their emission lines must require a compact central mass of about $10^8 M_\odot$.[13] Nevertheless, quasars were not immediately perceived as simply a more extreme version of the same phenomenon by most of the community.

Of the processes, accretion-powered x-rays were Hoyle's 1949 idea for explaining emission from the solar corona[14] (I am not sure we know the right answer to this one yet, though no one now believes that accretion is relevant), and synchrotron radiation was transported from the laboratory to the cosmos by Alfvén and Herlofson in 1950.[15] Just who was next to pick up the synchrotron torch to explain galactic radio radiation has caused some heated discussion, but the Soviet astrophysicists V. I. Ginzburg and I. S. Shklovsky were clearly involved one way or another.[16]

Gravitational lensing is normally thought of as a general relativistic phenomenon, but it is just as possible (to within a factor 2) with Newtonian gravity and a ballistic model of light. An early nineteenth-century note by J. Soldner on gravitational deflection of light was republished and discussed by Lenard shortly after the 1919 confirmation of general relativistic deflection by the sun.[17] My German is inadequate for judging just how prescient his remarks were. The early history of relativistic lensing is also somewhat confused. Einstein, who thought of one star lensing another,[18] and Zwicky, who thought more appropriately of one galaxy lensing another,[19] both credit the idea to R. W. Mandl (Zwicky through the intermediation of Zworykin of television fame). But a quarter-page note on lensing by Chwolson[20] appeared on the same journal page in 1924 as a brief note on another topic by Einstein,[21] and it seems to me virtually certain that Einstein must have looked at that page, if only to check that the editor had spelled his name correctly, and therefore must have seen the preceding note. A pair of still earlier short papers by Lodge[22] must also have been widely seen at the time. This is an area where some further investigation into who knew what and when might be fruitful.

While many undoubtedly recognized the need to analyze observations of radio sources with as much general relativistic sophistication as had been brought to bear on optical galaxies, G. C. McVittie[23] said it the loudest. Serious consideration of the early universe can be described somewhat similarly. The loudest early voice was surely that of George Gamow,[24] though the first estimate of a background radiation temperature came from Alpher and Hermann in 1949,[25] and the earliest basically correct calculation of primordial helium from Hayashi in 1950.[26]

There can be a considerable gap between the time an essentially correct idea is first proposed and the time it is generally accepted by the community. Books intended for use as texts provide an interesting snapshot of received opinion at particular epochs. In 1926, for instance, Eddington published his *Internal Constitution of the Stars*,[27] and Russell, Dugan, and Stewart their classic *Astronomy*.[28] One finds therein that the importance of general relativity for interpreting solar system observations, the gravitational redshifts of white dwarfs, and (probably) the large redshifts of the extragalactic nebulae was accepted, but stronger field effects were not. Eddington quotes Laplace and goes on to say that a star the size of Betelgeuse and the density of the sun "would produce so much curvature of the space–time metric that space would close up around the star, leaving us outside (i.e., nowhere)."

On the threshold of the relativistic astrophysics era, McLaughlin[29] and Baker[30] worry about the energy source replenishing magnetic field and relativistic electrons in the Crab Nebula but not about the energy for supernova explosions themselves. The source of galactic cosmic rays is "unknown," with acceleration tentatively attributed to the galactic magnetic field, and neutron stars are nowhere mentioned. They associate strong radio sources with galaxy collisions and are not prepared to accept the expansion of the universe without some reservation. And so forth. Only a decade later, Weinberg's book strikes one as essentially modern on these and many other topics.[31]

At the Symposia

Table 1 illustrates the gradual expansion of the terrain of relativistic astrophysics. It lists the 15 symposia in chronological order, with a few of the "hot topics" mentioned at each of them. The selection of topics has, unavoidably, been illuminated by hindsight, but I have tried to pick out items that occupied large sections of the program, large portions of my notebooks, or (at least in memory) large parts of the discussion time. The chief effect of hindsight is that most of the subjects tabulated are still regarded as important parts of relativistic astrophysics.

Some topics that once attracted a good deal of attention have gone out of fashion—occasionally because the problem has actually been solved, but more often because some other approach has captured most of the former adherents. Supposedly the human senses of smell and taste are the best memory provokers, but I find there is a good deal of nostalgia to be triggered by reciting lists like the following: $\log N - \log S$ and V/V_{max}; parametrized post-Newtonian approximation and exterior differential forms; white holes, antimatter, Hawking radiation, and mini black holes; Cf^{254}, the Dicke sun, Christmas tree models, and Geminga; spinors, twistors, fiber bundles, Kaluza–Klein, and minisuperspace; non-cosmological redshifts, $z = 1.95$, and $\Delta z = 0.061$; mixmaster universe, gravitational synchrotron, and tube of toothpaste effect. I suspect anyone who has had some interest in astronomy, astrophysics, or relativity for at least a decade will feel a slight twinge of "what ever happened to . . . " for at least a few of these.

Incidentally, there are at least a few people still working on each of these. Each time I have talked about this subject in public, one or two of them have jumped up at the end to say, "What do you mean, whatever happened to the mixmaster universe?! I published a paper on it last month." Fair enough; at some deep level I probably still believe that steady state is the right model of the universe, but it has gone out of fashion.

The Current Inventory

While the assortment of topics regarded as the core of relativistic astrophysics has not grown completely out of control, the various subfields of astronomy and astrophysics are rather closely interwoven. As a result, if a student were to ask what subjects he would have to know something about in order to understand all the talks at the next Texas

**TABLE 1. Dates, Locations, and Highlights
of the Texas Symposia**

WHEN	WHERE	WHAT
12/63	Dallas	quasars and gravitational collapse
12/65	Austin	3K background radiation; x-ray sources; supernova-cosmic-ray connection
1/67	New York	cosmological implications of radio source counts, angular diameters, nonclustering; radio-quiet quasars; Dicke sun and alternatives to GR
12/68	Dallas	pulsars
12/70	Austin	black and white holes; gravitational radiation
12/72	New York	x-ray binaries; accretion disks; superluminal motion in radio sources; solar neutrino deficit
12/74	Dallas	gamma-ray bursts; supernovae; binary pulsar
12/76	Boston	galactic evolution and q_0; black hole thermodynamics
12/78	Munich	quasar absorption lines; PeV/TeV gamma-ray sources
12/80	Baltimore	very large scale structure and deviation from Hubble flow; galaxy formation with neutrinos or explosions; Einstein x-ray data; SS433; gravitational lenses
12/82	Austin	particle physics and cosmology; cosmological nucleosynthesis
12/84	Jerusalem	strings; supernovae; cataclysmic variables; inflation
12/86	Chicago	supergravity; superstrings; cold dark matter
12/88	Dallas	supernova 1987A; low-mass x-ray binaries and binary pulsars
12/90	Brighton	particle astrophysics; nature and origin of large scale structure; early universe; neutron stars; ROSAT and COBE data

Symposium, one's initial response would be: everything from the Planck time to the great spots on Jupiter and Neptune. And the student would need to comprehend the observed properties of a good many astronomical objects as well as major branches of theoretical physics.

Table 2 is an attempt to classify "what you would have to know." It is divided among theoretical systems, processes, theoretical entities, and observed entities. Some of the things we talk about cannot quite be put firmly into either of the two latter categories. Supernova remnants, for instance, are fairly well categorized by their radio, optical, and x-ray emission properties, but, for all but a handful of cases, their birth in supernova explosions is an assumption. I should hasten to add that I fall very far short of understanding all the items in the table, and that my judgment of what to include is probably biased thereby.

Demographics and Literature

The People

The discoverers of the first quasars (see note 2) were scientists of European (including British) and American birth working in Australia and the United States. Hoyle and Fowler (see note 1) are, respectively, British and American. And the conveners of the first Texas Symposium were all part of the great World War II transplantation of scientists from the old world to the new (Bergmann, Robinson, and Schucking born in Europe, Schild in Turkey). Thus the discipline from the beginning was strongly concentrated in the United States and western Europe.

Soviet theoretical work on cosmology, gravitational collapse, and so forth was essentially contemporaneous, with three small, but very creative, groups centered around I. S. Shklovsky (an astronomer by training), V. I. Ginzburg (a physicist), and Ya. B. Zel'dovich (a chemical physicist with extraordinarily broad later interests). The first of these groups did early work on the Crab Nebula, quasar models, and accretion powering; the second on quasar models and radiation mechanisms, x-ray sources, and the Crab; and the third on collapse and accretion energy, gravitational radiation and neutrinos from collapsed objects, and cosmology. Although G. Zatsepin was at the 1st Texas symposium (and the 15th, as spokesman for the Soviet–American Gallium Experiment looking for solar neutrinos), it was not until the 3rd that Shklovsky and Ginzburg managed to get exit visas; and Zel'dovich never made it to Texas.

TABLE 2.　Entities Within Relativistic Astrophysics or Its Background

Theoretical systems

General relativity (etc.) and numerical relativity
Quantum electrodynamics (radiation mechanisms; particle acceleration)
Electroweak unification (neutral currents in supernovae, e.g.)
Grand unified theories (GUTs), supersymmetry (SUSY), supergravity (SUGR),
　　gauge theories, etc. (early universe)
Quantum cosmology and quantum gravitation

Compound theoretical systems

Particle dynamics and n-body simulations (classical and relativistic)
Plasma-, fluid-, hydro-, and magnetohydro-dynamics
Chaos
Stellar structure and evolution, single and binary
Galaxy formation and evolution
Cosmological nucleosynthesis
Density waves and solitons (spiral structure)
Core collapse and neutrino energy transport (supernovae)
Black hole electrodynamics (Hawking radiation, etc.)

Processes

Tests of general relativity
　　temporal changes or distance dependence of G
　　fifth forces
　　deflection of light, gravitational redshift, perihelion/periastron advance, time
　　　　dilation (first and second order)
　　dragging of inertial frames; geodetic precession
　　gravitational radiation
Tests of cosmological models: $H_0, \Lambda, q_0, \Omega$
　　counts of galaxies, radio sources, quasars
　　angular diameters
　　Hubble diagram (red shift *vs.* magnitude)
　　abundances of helium, deuterium, lithium
　　ages of globular clusters, white dwarfs, nucleocosmochronology
　　distance scale
　　mass to light ratios; dark matter

Theoretical entities

Stars supported by degenerate electron, neutron (etc.) pressure
Brown dwarfs, black holes

Pregalactic stars; supermassive objects
Accretion disks
Fermi acceleration
Nuclear detonation/deflagration
Relativistic star clusters
Phase transitions and inflation
WIMPs, inos, cosmions, lightest supersymmetric partner, axions . . .
Neutrinos of finite rest mass; MSW oscillations
Strings, cosmic and super

Objects observed but heavily theory-dependent

Supernova remnants
Intergalactic medium
Jets
Sachs–Wolfe effect; Sunyaev–Zel'dovich effect; Gunn–Peterson effect
Primordial perturbations; Harrison–Zel'dovich spectrum
Gravitational lenses
Gravitational radiation (binary pulsar)
Interacting and merging galaxies; galactic cannibalism
Superluminal motion

Observed objects

White dwarfs, cataclysmic variables, binary white dwarfs
Cosmic rays
Seyfert and radio galaxies, QSO's, quasars, BL Lac's, OVVs, etc.
X-ray binaries: high and low mass, black hole, bursting, transient, quasiperiodic
    oscillations, "pulsars," SS 433, globular cluster
Pulsars: strong field, binary, millisecond, globular cluster
Backgrounds at radio, millimeter, submillimeter, x-ray and gamma-ray energies
Solar neutrinos
Supernovae of types Ia, Ib, II (etc.?)
Neutrinos from SN 1987A
Solar oscillations
X rays from galaxies and clusters
QSO absorption lines: metal line systems, damped Lyman alpha, Lyman limit,
    Lyman alpha forest
Very large scale structure and deviations from Hubble flow: voids, superclusters,
    Rubin–Ford effect, great attractor, great wall
Gamma-ray sources: pulsed, erratic, bursting, Geminga
Galactic-center radio, x-ray, infrared sources; infrared lines and velocities
IRAS galaxies
PeV/TeV gamma rays

Australia and Canada were represented from the beginning by their radio astronomers and later by optical observers and theorists. The first Japanese involvement was primarily experimental. A set of 1964 review papers[32] reveals no fewer than four cosmic-ray groups, under S. Hayakawa, K. Sato, M. Oda, and M. Koshiba, with the first having then just recently branched out into x-ray astronomy. X-ray and computational astrophysics remain major areas of Japanese strength.

Chinese involvement in relativistic astrophysics has been primarily theoretical and rather limited. The participants at the 14th symposium voted overwhelmingly to send a special invitation for the 15th to Fang Lizhi (then not able to travel freely either within or outside China). He was indeed at the Brighton meeting, but as a resident of the United Kingdom.

One measure of how many people are interested and/or active in relativistic astrophysics is the number attending the biennial symposia. Figure 1 shows the best numbers I have been able to find (some from proceedings, some from notes of what the organizers said at the time, some from counting names on lists of participants distributed during registration). The pattern is very jagged, with highs and lows that undoubtedly depend heavily on the accessibility of the venues, the varying availability of travel funds, and the numbers of competing meetings. But the contrast with the exponential growth (5 to 9 percent per year) in numbers of members of the International Astronomical Union (Figure 2) is striking. Relativistic astrophysics would seem to have peaked in 1972 and to have been going down hill ever since (though the papers discussed in the next section tell a different story).

Figure 1 must, however, be interpreted in comparison with the participation patterns for other kinds of meetings. Membership in the American Astronomical Society, after some decades of exponential growth, has leveled off just above 5000 over the past five years. On the same time scale, gradually increasing attendance at the larger annual (January) meeting has topped out at about 1000 (± predictable venue-dependent fluctuations).

The IAU triennial general assembly, on the other hand, shows a pattern somewhat like Figure 1, with peaks and valleys. But the largest number of regularly registered participants came to Brighton in 1970 (2200 of them), and the peaks since have all been lower (apart from a very large number of local observer–participants at the Delhi 1985 General Assembly). IAU symposia seem to have saturated at 200 to 300 participants each, while the more specialized colloquia are probably still

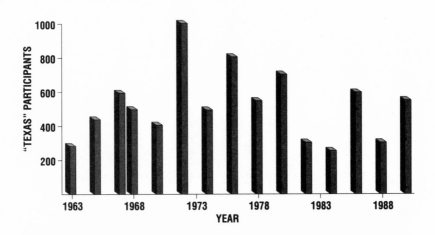

FIGURE 1. *Numbers of participants in the (roughly) biennial Texas Symposia on Relativistic Astrophysics. The highest peak is the 1972 New York meeting. Some numbers are rather approximate.*

getting larger and now sometimes exceed by an order of magnitude the 21 invited participants (and 2 gate crashers) of Colloquium number 6 in 1970, the peak general assembly year.

The most obvious interpretation of these patterns confirms what one's colleagues frequently say in private—that, increasingly, they prefer specialized meetings and workshops, preferably small ones, where "the people who are really working on the subject can talk to each other" and similar phrases. Unfortunately, the result of the preference being widespread is that there are really almost no small meetings left.

Another possible measure of interest in a field is society membership. There is no formal "Association of Relativistic Astrophysicists" (though the International Organizing Committee for the Texas Symposia has grown monotonically over the years). But the subject matter overlaps approximately with the areas of interest of (a) the Astrophysics Division of the American Physical Society, (b) the High Energy Astrophysics Division of the American Astronomical Society, and (c) Commissions 47 and 48 (Cosmology and High Energy Astrophysics) of the International Astronomical Union.

All three are about the same age. The APS division was founded in 1974 (as the Division of Cosmic Physics; other divisions go back to 1943),

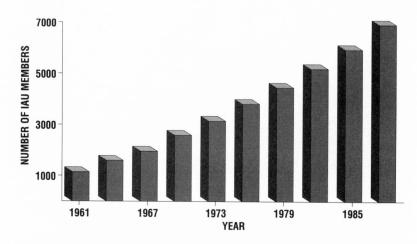

FIGURE 2. *Numbers of members of the International Astronomical Union at the beginnings of three-year intervals between general assemblies over the existence of relativistic astrophysics.*

primarily by cosmic-ray astronomers/physicists. The AAS division began as one of the three or so first topical divisions in 1974, with many x-ray astronomers among its members. And the IAU Commissions go back to 1970 (Commissions having been part of the initial structure of the Union in 1922). All three also include comparable fractions of the membership of their parent organizations. (Numbers are necessarily approximate owing to frequent additions and subtractions.) The APS division membership is 5.2 percent of about 41,000 APS members (1989 directory); the AAS division has about 7.2 percent of 5500 members (1990 directory); and the IAU Commissions include about 7.1 percent of the 7000 IAU members (after allowance for the people who belong to both). These percentages seem to have been fairly stable over the past decade, except that membership in all IAU commissions is rising faster than IAU membership in general.

The Literature

On the basis of both absolute and relative numbers of papers, one has to conclude that (contrary to the picture presented by participation in Texas Symposia) relativistic astrophysics is holding its own and even flourishing. By chance, *Annual Reviews of Astronomy and Astrophysics* also began in 1963. Figure 3 shows the fraction of reviews on topics in the area in four-year bins. Bin size was determined mostly by the possible ways of factoring 28 years. Following a sharp initial rise, that fraction has remained nearly constant around one third for a couple of decades. Owing entirely to remarkable discipline on the part of editors and publishers, the absolute number of papers has also been constant. The *Annual Reviews* included 128 papers in the eight years beginning in 1963 and also 128 in the eight years ending in 1990.

Paper length is another story. Despite continuing instructions from the editors concerning maximum number of permissible pages and a stable format and typeface, average review length rose from 26.8 pages in 1963–64 to 46.6 in 1989–90. This 73-percent increase over 25 years is typical of what has happened to the entire literature of astronomy, physics, and chemistry over the same period.[33] Table 3 lists authors and subjects of the reviews pertinent to relativistic astrophysics published during the periods 1963–67 and 1989–90. The chief perceptible trend is toward greater specialization.

Among research papers, both the absolute number and the fraction devoted to relativistic astrophysics have grown. *Astronomy and Astrophysics Abstracts* provides data from 1969.[34] (It succeeded a 70-year-old abstract series generally known as *Jahresbericht* whose subject categories make very difficult the identification and tabulation of papers on topics now regarded as part of relativistic astrophysics.) Although the *Abstracts* include essentially everything published in astronomy and astrophysics (book reviews, popularizations, reviews, and didactic papers, as well as conference proceedings, meeting abstracts, monographs, and journal papers), the dominant constituent is papers and abstracts reporting original research. Figure 4 shows the growth in numbers of authors producing papers in astronomy between 1963 and 1989. Some renormalization was needed between *Jahresbericht* and the *Abstracts*, because the latter surveys a wider range of journals and other publications. The first two points have, therefore, sizable error bars.

Between 1969 (when a total 10,800 papers were abstracted) and 1989 (with more than 20,000), relativistic astrophysics grew considerably faster than average, from about 12 percent of the total to 24 percent. Table 4 presents some of the details for all of 1969 and the second half of 1989. Subject classifications in the *Abstracts* have evolved over the years (the founders having presciently left spaces in their numbering scheme). This has been allowed for as well as I could by putting similar topics as close together in the table as possible. Notice that the subject inventory has not

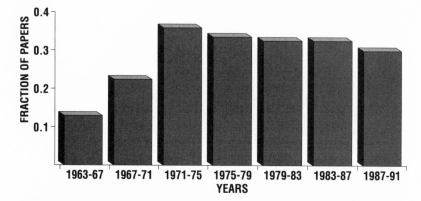

FIGURE 3. *Fractions of articles in* Annual Reviews of Astronomy and Astrophysics *addressing topics in relativistic astrophysics from 1963 to 1990 (four-year bins). Opening biographical chapters are included when the writer has done much of his work in the field.*

increased very much, so that the rapid growth has been accomplished by more work in relativistic astrophysics, not by previously existing work being relabeled.

Concluding Remarks

Since I began by describing this discussion as folklore of science rather than history, it seems appropriate to conclude with three anecdotes not previously published, so far as I know. Like all anecdotes, these are of considerably more interest to the teller than they are likely to be to any listener or reader. The first is set at the 1970 Brighton General Assembly of the International Astronomical Union where, as previously mentioned, the Cosmology and High Energy Astrophysics Commissions got their start. Owing to a shortage of small meeting rooms for such purposes, the organizing committee of the latter agreed to meet in the coffee room, "at Franco Pacini," to discuss putting their commission together, which they successfully did. Franco continues to play a large role in the field, both physically and intellectually.

The second tale I call "Gamow and the Graduate Student." It goes back to 1949, when a prospective graduate student, formerly a navy lieutenant commander working on electronic countermeasures, was looking

TABLE 3. Early and Recent *Annual Review* Articles in Relativistic Astrophysics

YEAR	AUTHOR(S)	SUBJECT
1963	C. Payne Gaposchkin	Novae
1964	R. Minkowski	Supernovae and supernova remnants
	B. Mills	Nonthermal galactic radio radiation
1965	G. Abell	Clustering of galaxies
	V. Ginzburg and S. Syrovatskii	Synchrotron radiation
	B. Burke	Galactic nucleus radio radiation
1966	J. Wheeler	Superdense stars
	A. Moffet	Structure of radio galaxies
	F. Gardner and J. Whiteoak	Polarization of cosmic radio waves
1967	G. Fazio	Gamma rays
	I. Novikov and Ya. B. Zel'dovich	Cosmology
	E. M. Burbidge	Quasistellar objects
	P. Morrison	Extrasolar x-rays
1989	L. Spitzer	Biographical notes (plasma astrophysics)
	G. Fabbiano	X rays from normal galaxies
	H. Bloemen	Diffuse galactic gamma-ray emission
	M. van der Klis	Quasi-periodic oscillations and noise in low-mass x-ray binaries
	W. D. Arnett *et al.*	Supernova 1987A
1990	V. Ginzburg	Biographical notes (cosmology; high-energy astrophysics)
	R. Canal, J. Isern, J. Labay	Origin of neutron stars in binary systems
	J. Higdon and R. Lingenfelter	Gamma-ray bursts
	F. D.A. Hartwick and D. Schade	Space distribution of quasars
	D. McCammon and W. Sanders	Soft x-ray background

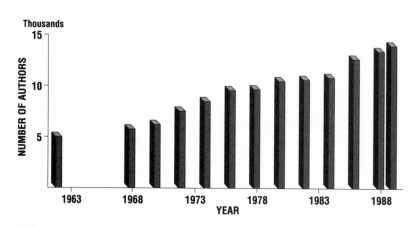

FIGURE 4. *Numbers of authors whose papers are listed in* Astronomy and Astrophysics Abstracts *(first two points are renormalized from* Jahresbericht, *which covered fewer journals). The number for each year is the number of authors in the larger of the two semiannual volumes and so represents roughly the number of people active enough to publish 1 to 2 papers per year.*

for a graduate program in physics convenient to his family and job in Maryland. Among the places he checked out was George Washington University, where George Gamow had landed. He found the right office and asked whether Gamow had in mind any interesting Ph.D. dissertation projects. "What can you do?" asked the professor. "I'm a microwave spectroscopist," replied the aspiring student. "No, I don't really think of anything at the moment," said the professor, thereby perhaps postponing by almost 15 years the discovery of the microwave background radiation. The necessary technology, including the Dicke switching radiometer, had been developed during the war. And I have enormous confidence in the engineering and research capabilities of the graduate student—my husband, Joseph Weber, who told me the story and draws the conclusion that Gamow did not then place much confidence in calculations of conditions in the early universe.

Finally, a vignette from my own graduate school days. A couple of years after Maarten Schmidt identified the first quasar, he found one with a red shift large enough that Lyman alpha (wavelength 1216 Å) appeared

TABLE 4. Numbers of Relativistic Astrophysics Papers
Listed in *Astronomy and Astrophysics Abstracts*, 1969 and 1989

TOPICS		PAPERS	
1969	**1989**	**1969** (all)	**1989** (second half)
radio sources, quasars,pulsars		470	
	radio sources		25
	quasars		114
	other active galaxies		379
cosmic rays		52	
	cosmic rays		67
white dwarfs		36	
	white dwarfs and pulsars		98
neutron stars and supernova remnants		46	
	neutron stars and black holes		259
	supernovae and supernova remnants		238
relativistic astrophysics, gravitation, 3K background		152	
	general relativity		362
theoretical cosmology		166	
	cosmology and background radiation		502
x-ray and gamma-ray sources		142	
	x rays		53
	gamma rays		56
x, gamma, neutrino processes; nucleosynthesis		97	
	nucleosynthesis, neutrinos, particles		164
MHD and plasma processes		105	
	plasma dynamics, magnetohydrodynamics		225
large scale structure, galaxy formation		13	
	lenses, IGM, clusters of galaxies		163
	TOTAL	1279	2705

PERCENT OF 10,800 (1969, all) 11.8%

PERCENT OF 11,079 (1989, second half)..................................24.4%

in the photographable part of the spectrum. A reproduction of this flashed on the screen early in a colloquium he gave at Caltech a few weeks later. Jim Gunn immediately declared to the other students around him in the audience (including Bruce Peterson and me), "That's important for cosmology." He had recognized, as others had not, that the presence of continuum emission at wavelengths shorter than that of the redshifted Lyman alpha line meant that there could be very little neutral hydrogen between us and the quasar in question. As a result, the universe cannot be closed by neutral or partially ionized gas. The constraint is now called the Gunn–Peterson effect and remains the most stringent limit we have on intergalactic gas at moderate temperatures.

If there is any lesson at all to be derived from these tales and the history of relativistic astrophysics in general, it is that the progress of science is exceedingly complex, nonmonotonic, and nonlinear.

ORIGINS

According to Hoyle

There is no longer really anything to be said about the original steady-state picture of the universe, in which matter is continuously created at just the right rate to keep all average properties constant for all time, except that it is a beautiful idea that turned out not to agree with observations. Hoyle has recently suggested an alternative picture, in which our universe changes with time, not because it is expanding but because the masses of the stable fundamental particles are increasing with time. Our universe is then just one of many patches in a space–time continuum, each with its own properties, which may have started its life with very large place-to-place fluctuations in the density of matter. S. E. Woosley (Caltech) has calculated the nuclear reactions that should occur early in such a universe and finds that a large fraction of the material in the densest clumps should be converted to elements with atomic numbers near that of iron. Conventional big-bang universes produce only hydrogen and helium early in their history. Besides providing an additional site for the synthesis of heavy elements, this picture has the attractive property that if the center of our sun is made of material processed in this way, the discrepancy between predicted and observed fluxes of solar neutrinos largely goes away. The chief difficulty may well be to avoid having all the iron so produced get trapped in black holes.

In more conventional cosmology, two of the classic tests, which it was once hoped would tell us whether our universe will expand forever or turn around and recontract, now no longer seem able to tell us this. The first of these tests is the relation between the red shift and the apparent brightness of galaxies. Since the apparent brightness of a galaxy decreases with its distance from us, and distant objects are seen as they were in the past, observations of the recession velocity (red shift) versus apparent brightness should tell us whether the expansion of the universe is slowing down enough for it eventually to stop and begin to contract. The number of available red shifts has increased considerably in the past few years (owing particularly to the work of Oke and Gunn at Hale Observatories), but the interpretation has become more difficult. In order to determine the distance to a galaxy from its apparent brightness, we need to know how its real brightness changes with time. It was once assumed that such changes were small, but this no longer seems to be the case. On the one hand, B. Tinsley (Yale) has shown that the general tendency of the aging of the stars in bright, massive galaxies is to make them grow significantly fainter with time, while on the other, J. P. Ostriker and S. D. Tremaine (Princeton) have pointed out that such galaxies will gobble up material from other nearby galaxies, which will make them grow brighter with time. It does not, at present, seem possible to decide which of the two effects dominates and thus to interpret the red shift–apparent brightness data.

The other classic test is to measure the apparent (angular) diameter of objects, whose real size you think you know, as a function of their distance. Because the structure of space–time (within the framework of general relativity) is determined by the amount of matter present, this test should also distinguish an ever-expanding universe from a re-contracting one. The optical data on this problem have always been difficult to interpret, but there had been great hopes for using radio sources. R. Ekers and others at the Westerbork Radio Observatory in the Netherlands have recently measured angular diameters of a large number of faint (hence distant) radio sources. Their data, in combination with previous results, say only that the real sizes and brightnesses of radio sources change with time. Since we have no theory of what the changes ought to be, we cannot now do cosmology with angular diameters.

Hoyle was among the first to predict the composition of the dust grains that pervade interstellar space. In the intervening years, his heretical suggestion—graphite—has become part of the conventional wisdom. N. C. Wickramasinghe (Cardiff) has proposed a new heresy, suggesting that polymers of formaldehyde and similar organic molecules may be important components, responsible for the observed ultraviolet absorption features. Even more controversial is the question of what fraction of the interstellar medium's supply of heavy elements is, in fact, locked up in the grains. Analysis of ultraviolet absorption lines in the spectra of distant stars, obtained with the Copernicus satellite, has normally been thought to imply considerable depletion of heavy elements from the gas onto the grains. G. Steigman (Yale) now points out, however, that most of the absorption probably takes place in ionized regions around the stars themselves, so the data tell us very little about the heavy element content of the general interstellar gas.

An increasing body of data may require us to rethink our ideas on the history of the solar system during the period when the planets and meteorites condensed and cooled. The traditional view requires a homogeneous gaseous medium to condense quickly and without outside interference. We now know of a variety of place-to-place variations in the ratios of isotope abundances of various elements in meteorites, which require some modification of this picture. D. D. Clayton (Rice) has suggested that interstellar grains condense in the immediate vicinity of the supernovae where the heavy elements are made and preserve their identity through the condensation process.

Are Galaxies Here to Stay?

WITH MARTIN REES

alaxies exist. Edwin P. Hubble settled this for most astronomers in 1923, when he identified Cepheid variables in the Andromeda and Triangulum spiral nebulae. He thereby established distances for the spirals that meant they must be enormous systems of stars rather like our own Milky Way galaxy. A typical galaxy contains some 200 billion visible stars. If we could collect a dollar from each star in our galaxy, it would pay off about half our national debt.

Not only do galaxies exist, but in one sense we even understand why they exist. The average density in our universe is so low that, in the absence of concentrations of matter like galaxies, there would be no chance for stars to form, give birth to planets, or make the chemical elements necessary for life. Thus no living creatures could ever have evolved, and we would not be here to worry about astronomical problems.

In another sense, however, our understanding of how galaxies form and reach their present attractive appearances is in a rather primitive state. In fact, things have become worse in the last few years. Instead of a single, coherent picture, with only details remaining to be sorted out, astronomers now have two seemingly contradictory scenarios, and only the beginnings of a synthesis between them. But that synthesis has some attractive features and may turn out to account for a large fraction of our observations of galaxies and clusters of galaxies without disagreeing with anything else we think we know. In other words, we may finally have hold of a reasonably good model of galaxy formation and evolution.

All the observations of galaxies that have been made in the past 50 years would (and do) fill hundreds of books. It would be unreasonable to expect one model, no matter how clever, to deal with them all. Thus the first step is to decide what features of galaxies are important enough to try to account for. These will normally be the properties that many galaxies have in common and the properties that vary systematically from one class of galaxy or cluster to another.

First on our list of important galaxy properties must be the basic shapes: elliptical, irregular, barred and normal spirals. But even here we can introduce some simplification. Spirals, for instance, consist of two components: a spheroidally shaped (or halo) distribution, consisting of old stars only, and a much flatter disc distribution, which includes the younger stars and gas. An elliptical galaxy, on the other hand, has only a spheroidal component, very much like that in the spirals. Irregular galaxies have many things in common with the more disorganized discs. In addition, there seem to be transition objects—galaxies which could be either irregulars or spirals with lots of disc and little halo, and lenticular galaxies, that could be either very flat ellipticals or spirals without much disc. What we really need to account for are two kinds of components—spheroidal and disclike—which can occur in different proportions to make different kinds of galaxies.

The two kinds of components behave differently. Discs usually have spiral arms, and the faster the disc rotates the more conspicuous and gassy the arms. Most of their mass is in the form of visible stars and gas, and none of the stars has metal abundances much less than 10 percent of the solar value.

Spheroidal components rotate slowly or not at all, and although they can be somewhat flattened, their basic structure is radial. Both the density of stars and their average metal content decline smoothly from high values at the center to low values 10,000 to 20,000 parsecs out into the halo. The outer parts of the halo are dominated by stars of very low mass (or something else that contributes a lot to the galaxy's gravity but hardly at all to its light). Far out in the halo, the metal abundance is always low, but it no longer varies systematically with the radius. Both spheroidal and disc components have systematic variations of color, mass-to-luminosity ratio, and other properties that are closely associated with the masses and compositions of their stars. For instance, other things being equal, the more massive a galaxy is, the redder it looks. The red color is not a result of red shifts but indicates relatively high metal abundance—suggesting that massive galaxies are more successful than small ones in retaining gas that has been processed through supernovae, where metals are made.

Galaxies "know" whether or not they are in clusters. Spirals, for example, are virtually absent from the middle of rich clusters. The gas has apparently been swept out of the galaxies and into the space between them. It can show up as part of a hot intracluster medium, whose x-ray emission reveals that the gas has an "almost solar" metal abundance. The central part of a cluster of galaxies also contains more than its fair share of the most massive galaxies—sometimes with an extrafat cD galaxy right at the center. A large fraction of a cluster's mass is hidden in the intracluster medium, probably in the form of low-mass stars, which have been stripped from the outer parts of galaxy halos.

Our understanding of galaxy formation and evolution is changing so fast that the "clever new approach" of 1976 is the "reasonably coherent picture" of 1977 and the "outmoded convention" of 1978. For instance, the hope that astronomers may be able to understand galactic evolution quantitatively is so recent that a spot check of the indexes of 18 post-1972 elementary astronomy texts (the traditional repository of the parts of science astronomers think they agree about at a given time) uncovers only 7 with any mention of at all of galactic evolution. Most of these merely say that nothing is known about it. Under these circumstances, it's a bit difficult to say what the conventional wisdom is. But if you'd asked a diligent reader of the literature and attender of astronomical conferences a couple of years ago to summarize what was known about the formation and evolution of galaxies, the result would have been some thing like the following:

"The very early universe may have been exceedingly chaotic, because light and other kinds of energy had not had time to travel from one place to another and smoothe things out. Chaotic temperature, density, and velocity variations were partially smoothed by diffusion of photons and neutrinos, but some lumpiness would survive. When the expanding universe cooled to a temperature near 3000 degrees Kelvin, protons and electrons combined to make neutral atoms. In that early era, the universe was a billion times denser than it is now—much denser than a present day galaxy or cluster.

"Although galaxies didn't then exist, they must have developed from lumps of above-average density. The lumps that could most readily survive the smoothing had characteristic masses comparable to those of present groups and clusters of galaxies (10^{12} to 10^{15} solar masses). When the material in such a surviving lump discovered that it was gravitationally bound, it stopped expanding with the rest of the universe and began to contract. Since this couldn't have happened until the average density of the universe had dropped to that in clusters now, contraction of clusters began about a billion years after the big-bang expansion started.

"The contracting surviving lumps were the protoclusters. Within them, some regions had higher densities than others and served as nuclei onto which more gas condensed. After another billion years or so, each sub-condensation had acquired most of the matter it would ever have and thus deserved the name protogalaxy. Gas that never made it into a proto-galaxy became the intracluster medium.

"What each protogalaxy did next depended on its mass, rotation, de-gree of turbulence, and perhaps some other things we don't know about yet. These in turn depended on the environment where the protogalaxy condensed, accounting for the fact that galaxies know where they are in clusters. All the protogalaxies were roughly spherical and began to make stars quite promptly. Among the first stars were some very massive ones that turned a few hundredths percent of the gas into metals and then mixed it throughout the rest of the gas.

"Star formation in the lumps continued, making the stars we now see in spheroidal distributions. If all the gas managed to condense at this ear-ly stage (presumably because the lump had a small amount of rotation and turbulence), the galaxy became an elliptical. The subsequent evolu-tion of ellipticals has been quite simple: Gas shed from the stars in super-novae, planetary nebulae, and so forth, was lost to the galaxy if its mass was small and its gravity weak. Dwarf ellipticals were left with a single generation of aging, metal-poor stars.

"More massive galaxies had enough gravity to keep their metal-rich gas. Their gas clouds collided, turning their energy of motion into heat which radiated away, allowing the gas to flow toward the center of the galaxy. As it flowed, it gradually condensed into new generations of metal-rich stars. Thus we see gradients of metal abundance, from low at the outside to high at the center, in the spheroidal components of galax-ies. Spectacular fireworks were likely to occur when the last of the gas arrived suddenly at the center and rapidly turned into many stars or a sin-gle supermassive object (or black hole). Some theorists hope (and their calculations even suggest) that the result will look like a quasar.

"Inside the more rapidly rotating and turbulent lumps, most of the gas did not turn into star immediately. Instead, the gas clouds converted their random motion into heat and radiation, delaying star formation. Their initial rotation caused these lumps to spin into flat discs. Star formation went rather slowly in these discs, and furthermore was largely confined to spiral arms generated in the disc by gravitational instabilities.

"These spiral galaxies have a halo population too, consisting of old stars and sharing the gradients and other properties of ellipticals. But they also have a flattened component, where stars have formed continu-ously from the time the remaining gas collapsed to a disc to the present."

Computerized versions of this model seem to be a tremendous success. They can account for many of the observed color, composition and brightness differences among galaxies (and between different places within galaxies), without feeding too many assumptions into the program. The do *not* deal very well with the random composition variations in galactic halos, with the earliest production of heavy elements in protogalaxies, or with the high metal abundance seen in intracluster gas. There are a couple of other problems. In this picture, galaxy formation and heavy element production (that is, intensive formation of bright young stars) occur late enough that we might expect to see very bright young galaxies at moderate distances. No evidence for these has been found. In addition, these models do not discuss the cases where there is strong evidence for direct gravitational interaction between galaxies, as in the centers of rich clusters.

In this model, galaxies are essentially closed systems. At most they may have some of their gas swept out into the cluster or receive a bit of gas that flows into a disc from a halo or intracluster medium late in life. But apart from this, galaxies and clusters will maintain their identities for most of the age of the universe.

Sir Isaac Newton solved the problem of two bodies moving under the force of their own gravity in several interesting cases, like the earth–moon system, the sun–earth system, and the earth–apple system. This can be done with a sharp pencil on the back of an envelope. The minute you introduce a third body, things get much more complicated and can only be solved in certain special cases. By the time you have a thousand or so bodies, the size of the required envelope back is roughly the capacity of the largest digital computers. Thus it is only quite recently that anyone has tackled the question: What happens if you plop down a bunch of small objects (for instance, small clusters of stars) in an expanding universe and let them all interact with each other via gravity?

The answer is a curious one: If someone will give you the clusters (of about 100 million solar masses each), then the formation of galaxies and clusters of galaxies takes care of itself. And the clusters made in the computer look very much like the ones we actually see in terms of numbers of members, where the massive galaxies are, and so forth.

Clearly, astronomers have the beginnings of a whole new way of trying to understand galaxy formation and evolution. Suppose that not long after the electrons and protons combined to form neutral hydrogen gas in the early universe, some of the lumps from the chaotic early universe began to fragment into stars. This could have happened long before the density of the universe had dropped low enough for protoclusters to form. And it has several advantages. These very early stars will pollute all the rest of the gas with metals, so we are no longer surprised that intracluster gas is metal-rich. When preexisting clusters interact to form galaxies, stars from a single cluster could end up almost anywhere in the new galaxy, so we are no longer surprised by composition variations in halos. And the interactions will continue right down to the present time, so that we expect to find galaxies in the process of capturing and assimilating each other, especially at the centers of rich clusters. And this is just where we find the overgrown cD galaxies.

But we have lost something. It's no longer easy to account for the systematic variations among galaxies and with position in spheroidal components that the gas-dynamical models dealt with so well.

By now you may well be way ahead of us, and already have asked the obvious next question: Since these two models are complementary, each dealing well with the phenomena that the other cannot account for, isn't there some way of combining the virtues of both? Clearly there is. We must consider both gas-dynamical and n-body processes.

The unified scenario starts, as usual, with a hot, dense early universe in which nuclear reactions turn about one quarter of the material into helium. Although the matter is cool enough for nuclei and electrons to combine and form ordinary atoms after about 10 million years, the formation of stars is inhibited for another 100 million years or so. During this period, even a few massive stars would ionize the gas enough to stop any further star formation. The first size scale on which anything interesting happens is the largest one on which there are big density fluctuations left from the early chaotic universe. Thus, clouds of about 100 million solar masses collapse and fragment into stars first.

About 90 percent of the gas turns into stars at this stage. The stars include some very massive ones that evolve rapidly, explode as supernovae, and throw the metals they have made into the remaining gas. But much of the fragmentation continues down to the minimum possible lump size—about 0.01 solar mass, depending somewhat on temperature. Such low-mass stars have long lifespans, longer than the present age of the universe, and are so faint that we could not see even the ones in the halo of our own galaxy. They may not even undergo nuclear reactions.

The 100 million solar mass clusters are then the units that interact gravitationally, collide, and eventually merge to form galaxy-sized lumps. Close encounters between pregalaxies can give some of them significant amounts of rotation. When the pregalaxies have built up to large enough masses, their gravity becomes great enough that the remaining gas falls into them. It does not immediately form stars, because every time a gas cloud comes too close to a star in the pregalaxy, it gets torn apart. Only when the gas has fallen in quite far, so that its average density is high, can it make stars and metals. Thus the parts of galaxies that we normally see are formed by this later infall, which acts very much like galaxy formation in the gas-dynamical model and produces abundance gradients, discs, and so forth. The invisible massive and very extensive halos surrounding the luminous parts of galaxies are made from the faint first-generation stars.

While the visible parts of galaxies are busily forming from accumulated gas, various gravitational processes continue to occur. They gradually build up small groups or galaxies (like our own Local Group) in which individual galaxies retain their massive halos, as well as rich clusters of galaxies, in which halos are stripped off and become the invisible diffuse mass of the intracluster medium. In rich clusters, gas will also be swept back out of galaxies quite early, accounting for the absence of spirals. As the galaxies continue to interact, the most massive ones accumulate near cluster centers. They eventually merge and "cannibalize" smaller neighbors to form giant cD galaxies.

Because these processes have acted over the entire history of the universe, the appearance of clusters and galaxies depends on when you look. If you had looked at the matter which now forms the Virgo cluster of galaxies when the universe was only a few hundred million years old, you would have seen characteristic units of about 1 billion small stars, quite different from the present 100-billion-star galaxies. And an observer 100 billion years in the future will see only a single, enormous cD galaxy there, with stars spilling out of it on all sides, and not much else but a few outlying dwarf galaxies. Hence, our "best buy" model indicates that galaxies are not a permanent feature of the universe, but are only characteristic of the time we happen to live in.

The "best buy" model appears to accommodate all the existing observations fairly well. Several additional tests should be possible in the future. For instance, in the gas-dynamical model, the protoclusters condensing some billions of years ago should introduce place-to-place variations in the microwave background radiation that has been knocking around the universe ever since the first few hot hours. But no such variations are to be expected if clusters are assembled from pre-existing stars and galaxies.

Similarly, hot protogalaxies or protoclusters would have emitted x rays which we would see as small "hot spots" on the sky. These would not be conspicuous according to the "best buy" model. Finally, if all the metals we now see were made after the galaxies formed, young galaxies, filled with hot young stars, must have been very bright. We should be able to see them—very distant, very redshifted—with suitable infrared devices in the next decade. No such bright phase is required in the combined model, since many of the metals were made before the galaxies existed.

Thus future observations by radio, x-ray, and infrared astronomers should be able to answer the question: Are galaxies here to stay?

Man's Place in the Universe

W hat I want to try to do in the next few pages is to review the history of the universe from the earliest times for which we have any evidence down to the present day, with special emphasis on how conditions favorable for life seem to have arisen, and then to explore the extent to which this history is dependent upon the universe having roughly the properties it does, and finally to inquire into the implications of varying those properties.

A Cook's Tour of the Universe and Its Early History

Let's start by taking a look (Table 1) at the scales of the things we will be discussing. Notice that the human scales in each case are close to the geometric means of the astronomical and atomic scales. Thus, we should not be surprised to find that our presence here is dependent both on the large scale phenomena of astronomy and on the details of atomic physics.

The largest phenomenon of all is, of course, the universe itself. It is important to be sure we agree about what we mean by "the universe" and the various other terms we will be using. The earth and eight other planets, about 34 moons, and a variety of smaller objects are in gravitationally bound orbits around a star called the sun. We refer to this grouping as the solar system. It has a total mass of about 2×10^{33} g (virtually all in the sun, though most of the angular momentum is in the planets), a diameter of about 2×10^{15} cm, and an age of about 5×10^9 years. The sun is a perfectly typical star, having a mass of 2×10^{33} g (the solar mass, abbreviated M_O, is often used as a unit for other stars), an electromagnetic radiation energy output of 4×10^{33} erg/sec (one solar luminosity, L_O), a spectrum approximately that of a 5700K black body, a radius of 7×10^{10} cm (1 R_O), and a composition by weight (at least in its outer visible lay-

TABLE 1. Scales of Phenomena Being Considered

Atomic scale	Human scale	Astronomical scale
TIME		
Nuclear decays 10^{-14} seconds	Attention span of physics undergraduates 1 minute = 60 seconds	Age of the universe 6×10^{17} seconds
MASS		
Hydrogen atom 2×10^{-24} grams	Typical Sigma Xi member 140 lb = 6.5×10^4 grams	Solar mass 2×10^{33} grams
LENGTH		
Diameter of atomic nucleus 10^{-13} cm	Height of dean at prestigious university 18 feet = 546 cm	Distance from sun to next star 1 parsec = 3×10^{18} cm
RATE OF ENERGY OUTPUT		
Atomic decay 10^{-3} erg/second	Output of large electricity generating plant 200 megawatts = 2×10^{15} ergs/second	Luminosity of the sun 4×10^{33} ergs/second

ers) of about 73 percent hydrogen, 25 percent helium, and 2 percent everything else (about half of it carbon and oxygen).

The sun, in turn, is one of about 2×10^{11} stars that are gravitationally bound in a rotating, roughly spherical system (although the most conspicuous members are concentrated in a plane considerably flatter than the proverbial pancake) called the Milky Way galaxy (or just the Galaxy). It has a mass of at least 3×10^{44} g (or perhaps as much as ten times this)[1] and a diameter of about 10^{23} cm. It is at least 10^{10} years old.

The Milky Way, in turn, is bound in a small cluster of about 30 galaxies (all but one much less massive than ours) called the Local Group. It is not certain whether higher-order structures are gravitationally bound, but there does seem to be some clustering of the clusters.[2] The clusters range from small ones like the Local Group to much richer ones containing thousands of galaxies and having masses of 10^{15} $M_\odot$. Completely isolated galaxies are probably rare.[3] The properties of the medium between the galaxies (except within the rich clusters, where hot intracluster gas is often a strong source of x rays[4]) are very poorly known. The average density could be anywhere from 0 to 10^{-5} particles cm^{-3}, the intergalactic medium comprising anywhere from 0 to 90 percent of the total average density over large regions of space. If the density is high, the matter must also be rather hot ($\sim 10^6$ K) or exceedingly clumpy to prevent detection. A preponderance of the evidence (as summarized, e.g., by Gott et al.[5]) now seems to favor an intergalactic density at the low end of the possible range.

The clusters of galaxies (or perhaps the superclusters) appear to be distributed at random through space, with separations such that they contribute an average density of at most 10^{-31} g/cm^{-3}.[6] There is no detectable falloff of the density of clusters of galaxies out to the largest distances at which they can be seen with present telescopes. This is about 10^{28} cm or 3000 Mpc (Megaparsecs; one parsec = distance at which an object has a *par*allax of one *sec*ond of arc = 3×10^{18} cm), corresponding to a light travel time of several billion years. We probably observe quasistellar objects (quasars) at much larger distances, but their properties are so poorly understood that they add very little to our knowledge of the large scale structure of the Universe.

The volume surveyed is sometimes called the observable universe, and it is the region for which we have direct observational evidence. Spectra of the vast majority of galaxies within this region and outside the Local Group show redshifts which are proportional to their distances from us. They are normally interpreted as Doppler shifts, implying that all the objects within the observable universe are receding from one another at speeds proportional to their separation. The proportionality constant is generally called the Hubble constant; its value (at the present time in the history of the universe and of astronomical research) is about 57 km/sec/Mpc.[7] This proportionality (Hubble's law) is our chief evidence that the universe is expanding.

Within the framework of some reasonable theory of gravity, like general relativity, we can extrapolate beyond the observable region and try to learn something about the entire four-dimensional space-time volume that can, in principle, be connected to us by light signals. The word *universe* properly refers to this entire volume and, in this sense, is not much more than 60 years old, dating back to the realization that certain bright, fuzzy patches in the sky are, in fact, galaxies like our own.[8]

Efforts to model the universe go back just about as far and always involve a variety of simplifying assumptions. The simplest possible set of such assumptions has proved remarkably successful. We assume (1) that general relativity is the right theory of gravity (probably without the arbitrary additional repulsive kind of gravity, the cosmological constant, introduced by Einstein to permit a static universe), (2) that the expansion implied by Hubble's law is isotropic and would be seen to be isotropic by any observer moving with the galaxies, (3) that the universe is homogeneous on a sufficiently large scale, and (4) that pressure is at present unimportant and that the mass-energy of the universe is now mostly in the form of matter rather than radiation or other zero-rest-mass particles.

Under these assumptions, the Einstein field equations yield a two-parameter family of models, called the Friedmann models, and the problem of deciding what the universe is like reduces to finding values for the two parameters. These turn out to be the present value of the Hubble constant, H_0, which we know to a factor of two, and the present value of the local average density of mass energy (in all forms), ρ_0, which may be uncertain by a factor of 100. Given values of these, we can answer a variety of interesting questions, like: How old is the universe? Is it finite or infinite in volume? Will the present expansion continue forever or will gravity cause the galaxies to slow down and eventually fall back together? Roughly, a low-density ($\rho_0 < 10^{-30}$ g/cm^{-3}) universe has infinite volume, is 16 to 20 billion years old, and will expand forever, while a high-density universe ($\rho_0 > 10^{-30}$ g/cm^{-3}) has finite volume, is less than 16 billion years old, and will eventually (in a hundred billion years or so) turn around and recontract. With many ifs, buts, maybes, and other caveats, evidence now available seems to indicate that our universe is a low-density one.[9] Notice that if the universe is to be a high-density one, then ≥ 90 percent of the mass-energy is neither visible nor in galaxies.

Under the same assumptions, there are some questions that we cannot answer or even ask in a meaningful fashion. One of these is Where is the center of the universe? The assumed homogeneity and isotropy of the expansion imply that all the matter was arbitrarily close together a finite time ago in the past, so that the center of the expansion exists only in four dimensions and is something like the instant of creation, while the geometry of space–time within the framework of general relativity is such that space is either infinite (and so can have no center) or uniformly curved, so that all points are equivalent (rather like the curved surface of the earth, only in three dimensions). It will become clear shortly that "What came before the present universe?" is another of these unanswerable questions.

With this background, we can now say that the earliest event for which we have any evidence is a time about 15 to 20 billion years ago when the universe was much hotter and denser than it is at present. The evidence for the time scale comes from (1) running the Hubble expansion backwards in time and asking how long ago would all of the galaxies have been arbitrarily close together (H_0 = 50 km/sec/Mpc = 1.67×10^{-18} sec^{-1}, or $1/H_0 = 2 \times 10^{10}$ years), (2) the ages of the oldest stars in our galaxy, probably 12 to 18 billion years, and (3) the ages of the radioactive elements in the solar system, which tell us that the earth and meteorites solidified about 4.65 billion years ago and that synthesis of these elements had been going on for 7 to 13 billion years before that.[10] The evidence for the high temperature and density comes (1) again from running the Hubble expansion backwards, conserving mass-energy and the numbers of various kinds of particles, including photons, and (2) from the existence of two relics of the hot, dense phase. These relics are an isotropic background of microwaves having a black-body spectrum corresponding to a present temperature of 2.7 K[11] and the seemingly universal presence of helium with an abundance of 20 to 30 percent by weight.[12] Thus, if we run a motion picture of the universe backwards about 20 billion years, we see it at a temperature $T \gtrsim 10^{10}$ K and a density $\rho \gtrsim 1$ g/cm^{-3}. Under these circumstances, many kinds of matter and radiation come into equilibrium, and the relative numbers of various kinds of particles (protons, neutrons, electrons, positrons, neutrinos, photons, and perhaps others) depend only on T. As the universe expands and cools, unstable particles decay or annihilate; and others undergo nuclear reactions, resulting in about 25 percent He^4 and traces of H^2 (deuterium), He^3, and Li^7, as well as about 75 percent ordinary hydrogen in the standard cosmological models.[13]

Unfortunately, this hot, dense phase (sometimes called the Big Bang) also wiped out any evidence of what (if anything) went on before. Hence the question "What happened before the Big Bang?" belongs to the realm of pure speculation (philosophy?) rather than that of physics. It is rather like putting a car into a steel blast furnace and asking the trickle of molten metal that comes out whether it was a Pinto or a VW before. You just can't tell, because the evidence has been destroyed.

Galaxies and Stars

Coming out of the hot, dense early universe we therefore see some radiation (which continues to cool, down to 3 K by the present time) and some matter, in the form of hydrogen and helium. Luckily this is not all that happened, because the chemistry of H and He is not very interesting. The matter at this stage was not perfectly smooth but was concentrated in lumps. This is also fortunate for us, because, as we have already seen, the average density of matter in the universe is exceedingly low. Thus in the absence of local concentrations of matter, the average hydrogen atom would not have encountered another hydrogen atom for the last 10 billion years or so, and would be very lonely. The cause of the lumps is not well understood, though they are not unexpected, since, when the universe was very young, there had not yet been time for interactions and smoothing to have occurred across large distances. But the lumps must have been there, because we see galaxies and clusters now. There has been some success in calculating how the lumps must have grown and developed into protogalaxies.[14]

The evolution of a galaxy is largely a matter of the exchange of material between stars and an interstellar gaseous medium and the nuclear processes that occur in stars. Many different types of galaxies are observed (of which the most clearly defined are Ellipticals, with their brightest stars distributed throughout a spheroidal volume, and Spirals, with their brightest stars concentrated in spiral-shaped arms in a plane; see any elementary astronomy textbook for pretty pictures). They come with different total masses, luminosities, kinds of stars, colors, amounts of gas, shapes, abundances of heavy elements, and spatial distributions of these and other properties. Serious efforts to understand the evolution of galaxies go back only to about 1967, and the field is changing so rapidly now that it is hard to say more than that it looks as if we may be able to account for the observed ranges of properties and their correlations in terms of a rather small number of initial conditions in the lumps, e.g., total mass, size, angular momentum, and degree of turbulence.[15]

A protogalaxy becomes a galaxy when some appreciable fraction of its gas has been converted into stars. The most distant galaxies, seen as they were 3 to 5 billion years ago, do not look very different from the nearby ones, but the quasars may represent some early stage of the galaxy formation process, which appears to have been largely completed within a few billion years after the initial hot, dense phase of the universe. We do not, in other words, see any obviously young galaxies nearby.

The process of star formation from interstellar gas, on the other hand, has continued to the present time in our own and most other galaxies (though at widely varying rates). We can almost see it happen before our eyes, at least in the sense that we see some bright, naked-eye stars (for instance the brightest ones in Orion) that did not exist as separate bodies at the time of our remote Zinjanthropan ancestors. We do not have an adequate theory of star formation (lots of people would say we have none at all), but we can learn quite a lot about it by looking.

In our region of the galaxy (often called the solar neighborhood), young stars are almost invariably found in groups and in the presence of denser-than-average clouds of gas and dust in the spiral arms of the galaxy. The very youngest stars are typically still hidden behind the remnants of the clouds from which they formed and are seen only as infrared sources. We are led, then, to a picture in which a typical dense (10^{3-5} cm^{-3}), massive ($\sim 10^5$ M$_O$) interstellar cloud (these are observed as sources of radio line emission) is shocked by a collision with another cloud, with the expanding shell from one of the supernovae (which we will meet later), or with the density wave believed to be responsible for the characteristic spiral shape of our own and many other galaxies. The shock starts the cloud collapsing under its own weight. As it contracts, excess angular momentum forces it to fragment into star-sized pieces. This means masses from about 0.08 to 100 M$_O$, the lower limit being set by the requirement that the center of the piece eventually become hot enough for nuclear reactions to occur and the upper limit by the tendency for radiation pressure to blow material off a star that is too massive and bright. We cannot predict how many stars of each mass will form from a particular cloud (this is one of the senses in which we have no theory of star formation), but observation shows that, over the history of our own and most other galaxies, more than half the mass has gone into stars less massive than the sun. Thus small fragments are favored over large ones, and the smallest stars are the most common.

The fragments are called protostars and continue to contract under their own self-gravity (on a time scale that depends on their mass, amounting to about 0.1 percent of the time they will spend undergoing nuclear reactions) until their centers reach a temperature of about 10^7 K. Because the very first generation of stars must have been nearly pure hydrogen and helium, a variety of processes that now occur during the contraction phase (including, probably, those that lead to the formation of planets) will not have happened. At a central temperature near 10^7 K, hydrogen begins to fuse to form helium (either directly, or using carbon, nitrogen, and oxygen as catalysts, when they are available and at slightly higher temperature) with the liberation of about 7×10^{18} ergs/g.

Stars have clearly solved the problem of controlled nuclear fusion, which we are now attacking in the laboratory. The differences are (1) the stars can start with pure H and convert two protons to neutrons (via the weak interaction) to make He, while we must start with substances (deuterium, tritium, lithium) in which this has already been done; (2) the star does it at much higher temperature and density than is contemplated in the lab; and (3) the star will take 10^{6-12} years to get all the energy out, while we hope for rather faster results.

The energy liberated by hydrogen burning keeps the pressure inside the star constant and stops its contraction. The star will remain in hydrostatic equilibrium until its hydrogen fuel is exhausted. The hydrogen-burning phase of stellar life is called the Main Sequence stage, from the star's position on a diagram of brightness vs. surface temperature (Figures 1 and 2). Such plots are called Hertzsprung–Russell (HR) diagrams and are of considerable assistance in understanding stellar evolution. Notice the one-to-one relationship between stellar mass and position on the Main Sequence. This, in turn, implies a relation between mass and life span, since the fuel supply depends on mass and the rate of fuel consumption depends on luminosity (which scales about M^3 on the Main Sequence, as was first understood by Eddington).[16] A star's structure remains stable until hydrogen has been exhausted in about the inner 10 percent of its mass. This implies a Main Sequence lifetime of nearly 10^{10} years for a star like the sun (70 percent hydrogen in the inner 2×10^{32} grams, yielding 7×10^{18} ergs/g, and being used to supply 4×10^{33} ergs/sec). But a 20 to 100 $M_\odot$ star at the upper end of the Main Sequence can last only 10^{6-7} years, while stars of less than about 0.75 $M_\odot$ have not had time to leave the Main Sequence in the age of the universe. Almost 90 percent of the stars we see are on the Main Sequence, accounting for the name and implying that it is the longest-lived phase.

Most astronomers and physicists feel that we have a good quantitative understanding of the Main Sequence phase (despite the continuing deficit of solar neutrinos[17]), but the ratio of speculation to "hard" theoretical (and observational) facts will gradually increase as we move away from the main sequence. The evolution of normal stars as a function of mass has been studied by numerous groups. Three series of papers by W. D. Arnett, I. Iben, and B. Paczyński cover many of the important evolutionary phases.[18]

Exhaustion of hydrogen in the stellar core introduces a discontinuity in composition and mean molecular weight. As a result, the equilibrium structure becomes a very extended one.[19] The outer layers of the star rapidly expand and cool, while the core once again contracts under its own weight, liberating gravitational potential energy to keep the star

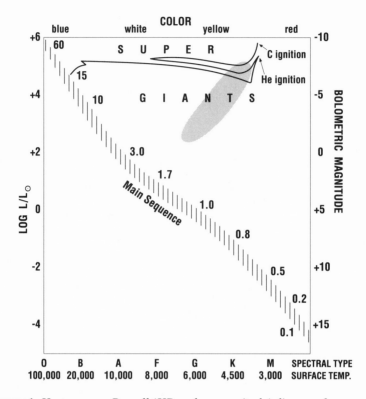

FIGURE 1. *Hertzsprung–Russell (HR; color-magnitude) diagram for a representative group of stars young enough that the entire Main Sequence is still populated. Notice the curious scales always used by astronomers. The vertical scale is the logarithm of total luminosity in solar units or magnitudes, which are logarithms (base $100^{1/5}$) of the reciprocal of the luminosity in fairly arbitrary units. The horizontal scale is surface temperature (though neither exactly linear or logarithmic) or color or spectral class (an ancient and honorable way of dividing up the stars that somewhat predates the realization that temperature is the most important determinant of spectral line intensities) and runs backwards. Masses in solar masses are given at representative points along the main sequence. The evolution of a typical massive star through the supergiant region is shown, along with the points at which helium and carbon burning start. Most of the time is spent close to the main sequence and in the red supergiant region. The stars in the stippled region of the diagram are generally or always variable in luminosity with regular periods of 3 to 30 days. They are called Cepheid variables, after the prototype, Delta Cephei, and are important distance indicators for nearby galaxies because their periods are correlated with their total luminosities.*

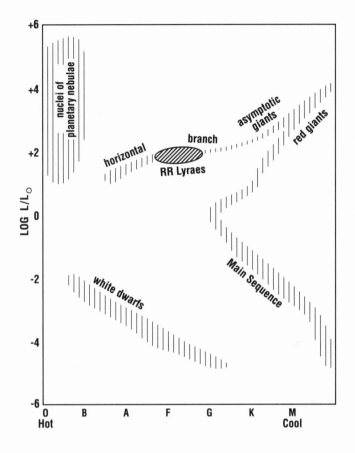

FIGURE 2. *Hertzsprung–Russell diagram for a cluster of stars old enough* ($\sim 10^{10}$ yr) *that stars like the sun* ($\sim 1M_{\odot}$) *are leaving the Main Sequence. The evolutionary track of a single star would go from the Main Sequence up into the red giant region along the heavily populated track, down to the horizontal branch (when He is ignited), back up the red giant region along the asymptotic branch, then horizontally (and very rapidly) across from right to left as a planetary nebula is shed, and finally down into the white dwarf region. The stars in the region labeled "RR Lyraes" are variables (named for their prototype RR Lyrae) with periods less than about a day.*

shining. The resulting star is called a red giant (or red supergiant, in the case of the most massive stars, which become even bigger and brighter) for obvious reasons.

Our sun will become a red giant in about another five billion years. When it does, it may become so large that its outer layers engulf the earth. In any case, its greatly increased luminosity is expected to raise the temperature of the earth to the point where the oceans boil away.

The increasing temperature of the contracting stellar core soon raises the temperature of the surrounding hydrogen enough that it begins to fuse to helium, again liberating nuclear energy. About as much hydrogen is burned in the red giant phase as was burned on the Main Sequence, but the star is about 10 times as bright, so the phase lasts only about 10% as long. Red giants (Betelguese and Antares are examples) are thus rarer in the sky than Main Sequence stars.

For all but very tiny stars ($\lesssim 0.3$ M$_\odot$, which have not yet had time to leave the Main Sequence anyway), the core eventually becomes hot enough for nuclear reactions involving helium to occur. The onset of helium burning occurs explosively in stars like the sun (because the cores are so dense the matter is partly degenerate; thus pressure does not immediately increase when the reactions drive the temperature up) causing a readjustment of the stellar structure. The star lands in the horizontal branch region of the HR diagram (Figure 2), while more massive stars merely remain in the red giant region during helium burning (Figure 1). Both phenomena are seen—the lower-mass helium burners as a collection of stars strung out horizontally across the HR diagrams of old star clusters, in which stars ~1 M$_\odot$ are leaving the Main Sequence, and the high-mass helium burners as a clump on the red giant branch of the HR diagrams of young clusters, in which ~2 to 10 M$_\odot$ stars are now leaving the Main Sequence. The helium burning phase is still shorter than the red giant phase, because the star remains quite bright and helium burning produces fewer ergs per gram than hydrogen burning.

The products of helium burning are carbon and oxygen, in roughly equal amounts. This is clearly of some importance, since we and other terrestrial living creatures are in large measure made of them. The approximately equal production of carbon and oxygen is the result of a delicate balance between the electromagnetic repulsion of the interacting helium atoms and the detailed nuclear structure of the products[20] and, in the absence of some deeper understanding of the forces involved, must be regarded as an extremely fortunate coincidence from our point of view.

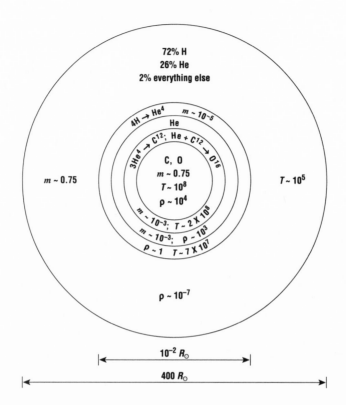

FIGURE 3. *Interior structure of a 1.5 $M_\odot$ (solar-type) star shortly before it sheds its planetary nebula and becomes a white dwarf. Masses of the various zones are given in solar masses, temperatures in Kelvins, and densities of the zones in g/cm³. Only the primary constituents and nuclear reactions are indicated. Notice the great disparity in scale between the inner and outer regions.*

The exhaustion of helium at the center of the star, like that of hydrogen earlier, results in the core contracting to liberate gravitational potential energy and the outer layers again expanding. Helium soon begins to burn a thin shell around the inert carbon–oxygen core (Figures 3 and 4). During this double-shell-burning phase, low-mass stars again ascend to the red giant region of the HR diagram (where, in old clusters, we see them as a scattering of stars to the blue side of the normal red giant branch, the so-called asymptotic giant branch), more massive ones remaining as red supergiants. During this phase, convection can bring the

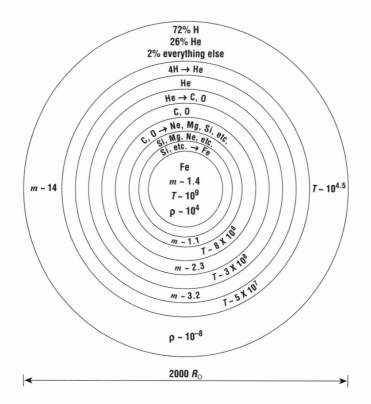

FIGURE 4. *Interior structure of a massive (22 $M_\odot$) star shortly before it becomes a supernova. Only the primary constituents are shown, and those rather schematically. Masses of the various zones are in solar masses, temperatures in Kelvins, and densities in g/cm^{-3}. Notice again that more than half the mass and virtually all the volume are still occupied by the hydrogen-rich envelope, whose composition has been relatively unaffected by nuclear processing in the star.*

products of nuclear reactions in the stars themselves up to their surfaces. Many asymptotic giant stars show excess carbon (and *s*-process products) in their atmospheres. In addition, in second and later generation stars, which had some heavy elements to begin with, several reactions involving He, C, N, and Ne liberate some neutrons. These are captured, primarily by iron and heavier nuclei, building up many (but not all) of the elements and isotopes between iron and lead (atomic numbers 27 to 82); by the so-called *s*-process (the slow addition of neutrons, interspersed with beta decays).

From here on, the evolution of a star depends mostly on its mass, following one path for stars $\lesssim 6\ M_O$ and another for stars $\gtrsim 6\ M_O$. Near the boundary, carbon burning may start so explosively (again because the core is degenerate) that the star is completely disrupted. The precise value of the mass cut between the paths has been discussed often and from many viewpoints.[21]

All stars shed some material throughout their lives. We see a continuous solar wind of particles leaving our sun, and the rate of mass loss is observed to be much larger in some bright red stars. Thus a star may reach the end of helium-shell burning with only half to three quarters of its original mass (or even less, if it is in a binary system). But it is the original mass that counts, the center of the star continuing to evolve almost unaware of the surface losses. The lower-mass stars, like our sun, never get hot enough for any further nuclear reactions to occur after helium burning. Rather, various instabilities in the shells gradually lift off the outer layers of the star, leaving behind the very hot, very dense core. The lost layers are heated and ionized by light from the remaining core and become responsible for one of the more frequently photographed phenomena of astronomy, the planetary nebulae (again, see any elementary text for pictures). The name reflects their appearance in a small telescope, not any imagined connection with planets.

The remaining core has a density of about 10^6 cm^{-3} and a radius of only about 10^4 km (the size of the earth). It is initially very hot even at the surface (10^{5-6} K) since it has just finished nuclear reactions, sometimes resulting in a detectable source of soft x-rays,[22] but cools off in a matter of millions of years to some tens of thousands of degrees. The nebulae dissipate within 10^4 years (judging from their measured expansion rates), leaving behind faint (because they are tiny) stars, moving in the HR diagram from the region labeled "nuclei of planetary nebulae" down toward the white dwarf region. The matter in these stellar cores is degenerate; thus they can neither radiate away the kinetic energy of the electrons nor contract any further. Degenerate matter is subject to the uncertainty principle; thus to compress it you must localize the particles further, raising the uncertainties in their momenta and, therefore, their kinetic energies. You must add energy to degenerate matter in order to compress it. Only the thermal energy of the nuclei is available and is gradually radiated away over billions of years, leaving the star as a gradually cooler and fainter white (and eventually yellow, red, then black) dwarf. This will be the final state of our sun. As seen from earth (which will, of course, be thoroughly frozen), it will not look much brighter than the full moon does now. The answer to Robert Frost's question therefore appears to be fire first and ice later.

Notice that the white dwarfs retain most or all of the carbon and oxygen produced in low-mass stars. Thus, we are going to have to follow the evolution of more massive stars to see the origins of the heavy elements found in the sun, the earth, and ourselves. In stars above about 6 M_O, the carbon–oxygen core gets hot enough for further nuclear reactions to occur. The burning of carbon and oxygen produces relatively abundant elements like neon, magnesium, silicon, and sulfur. As one source of fuel is exhausted, further contraction of the core causes further heating until another fuel can be tapped. Finally, the burning of elements near silicon produces a stellar core made mostly of iron and nickel, the star as a whole rather resembling an onion (see Figure 4). Energy is liberated in each of the nuclear reactions, though not as much as in hydrogen or helium burning, but most of it is carried away by neutrinos (rather than photons), which do not contribute to keeping the star shining. These phases are therefore very short lived ($<10^4$ years), and the HR diagram is no longer a particularly useful tool for following the evolution.

Once the star develops an iron core of about 1 M_O, it is in serious trouble. The nucleus of iron is the most tightly bound of all atomic nuclei; hence no reaction involving it can liberate any energy. The core of the star just continues to contract and get hotter and denser until several processes begin to occur (more or less simultaneously) which drain energy away from the core, causing it to collapse suddenly and catastrophically. The two main processes are (1) the breaking up of Fe back into He nuclei by high-energy gamma rays and (2) the reaction of protons with electrons to form neutrons and neutrinos. Both of these soak up energy, and the latter especially also reduces the effective sizes of the particles in the stellar core. The core therefore collapses until a new source of pressure (the neutrons becoming degenerate) stops it at a radius of about 10 km. The product is called a neutron star, and the process liberates some 10^{53} ergs of gravitational potential energy (comparable with the total radiation output of the star over its entire previous life).

This liberation of energy and its distribution through the star result in a variety of violent phenomena, including (probably) (1) a burst of neutrinos, (2) a burst of gravitons, (3) the *r*apid addition (*r*-process) of neutrons to iron from the core to build heavy elements up to at least plutonium and maybe further, (4) some very high-temperature nuclear reactions in the intermediate layers which yield a variety of rare, intermediate-weight isotopes, (5) a rapid brightening of outer layers of the star until it may outshine its entire galaxy for some weeks (we call this a supernova when we see it, which happens once every few hundred years in the solar neighborhood, or about every 30 years in a large galaxy), (6) the acceleration of some material from the surface of the star to relativistic speeds, after which we call the particles cosmic rays, (7) the expulsion of one to many solar masses of material from the star at speeds of 10^{3-4} km/sec (we see such expanding clouds around old supernovae in the solar neighborhood), and (8) the storing of energy in rapid rotation of the neutron star, which then gradually feeds it out again over the next 10^{6-7} years in forms that cause us to see it as a pulsar. The association of neutron star formation with supernovae was first suggested by Baade and Zwicky[23] and confirmed by the detection of a pulsar in the Crab Nebula,[24] the gaseous remnant around the site where the Chinese reported seeing a supernova explosion in 1054 C.E. From our point of view, the most interesting thing a supernova does is to distribute the heavy elements made by nuclear reactions in the massive star back into the interstellar medium.

I have said nothing so far about the evolution of pairs of stars—binary systems—which constitute at least half the stars in the solar neighborhood. The presence of a close companion influences a star's life profoundly,[25] particularly in the later stages. It may even lead to an occasional massive star collapsing past the neutron star stage down inside its own Schwarzschild horizon to form a black hole.[26] No one has yet followed a massive star in a close binary system all the way from the main sequence to supernova explosion, and the effects on nucleosynthesis are unknown but thought to be small. The core of the star should continue on its merry way even after the outer layers are stripped off and given to the companion star.

In the case of normal supernovae, we observe the shedding of the outer layers, and careful spectroscopic studies have found extra amounts of various heavy elements in them,[27] representing the products of nuclear reactions in the stars. These then enrich the general interstellar medium, so that second and subsequent generations of stars that form will contain a component of elements besides hydrogen and helium (about 2 percent in our sun and 3 to 4 percent in stars now forming), in proportions that are reasonably well understood theoretically.[28] The presence of heavy elements in turn implies the possibility of planets like the earth.

Planets and Life

If we are interested in life in the universe, one of the first things we will ask a star is whether or not it has planets, preferably with solid surfaces and preferably located at such a distance from the parent star that water is a liquid at least sometimes and in some places. Unfortunately, direct observation cannot answer this question for any star but the sun. In the past, theories of the formation of the solar system often involved events, like the close passage of two stars, so rare that we could easily be the only planetary system in the galaxy. Recent theoretical studies have, however, come around to a view in which leaving behind some material in planets is a natural part of condensing a star out of the interstellar medium. On this basis, planets should be quite common. There is some indirect supporting evidence.[29] Careful study of the motions of several nearby stars across the plane of the sky has revealed evidence of a roughly sinusoidal wiggle, typically with a period of years and an amplitude of a small fraction of a second of arc, such as would be produced by the star and a planet with a mass 1 to 10 times that of Jupiter orbiting around their mutual center of mass. And if the presence of Jupiters can be taken to imply the presence of Earths, then many or most stars (at least those of roughly solar type) may well have planets with solid surfaces.

The nature of the atmospheres of these planets will also be important for the possibility of life. The earth's original atmosphere was not its present oxidizing one, but rather a reducing one (made up of things like CO, CO_2, NH_3, CH_4, H_2, and H_2O at various stages) according to several lines of evidence. These include the kinds of gases found on other planets and released by volcanoes, the reduced nature of the minerals deposited until about a billion years ago, and the necessary conditions for the origin of life itself.

The constituents of this primitive atmosphere are interesting from several points of view. First, they are made of the commonest chemically reacting elements (H, O, C, and N, in order of abundance). Second, they contain the elements that occur most abundantly in terrestrial living creatures. Third, they are more or less the things you need in order to do the kind of experiment suggested by Urey and carried out by Miller[30] (and many others since) in which substances like methane, ammonia, formaldehyde, and carbon dioxide are dissolved in water and energy added (in the form of ultraviolet radiation, electric currents, mechanical shocks, and so on). Under these circumstances, many chemical reactions occur, whose products (if the right inorganic reactants are made available) include a wide variety of simple organic molecules—amino acids, sugars, bases, phosphates—and some of their polymers.[31] And fourth, these and a wide variety of other simple organic and inorganic molecules are now known to be widely distributed through the interstellar medium.[32] Molecules detected include interstellar CO, CN, NH_3, H_2S, H_2O, formaldehyde and hydrocyanic acid, formic acid, methyl and ethyl alcohol, and HC_5N, which has the same molecular weight as glycine, the simplest amino acid, and about 30 others. These molecules and the presence of amino acids (in a racemic mixture of left- and right-handed forms, which leads us to believe they are nonbiological) in several meteorites[33] seem to indicate that both the raw materials and pre-biological organic molecules will be widely distributed in suitable environments, like the surfaces of planets.

No laboratory experiment has yet taken the complete step from the raw materials to a self-reproducing molecule. Perhaps this is just as well; the first biochemist who makes a self-reproducing molecule may find out that it eats biochemists. Ways to approach self-reproduction have been discussed by Calvin.[34] In any case we (or at least the infidels among us) do know that the step has been taken at least once, for here we are; and, by the time you read this, laboratory experiments, planetary probes, or searches for messages may have shown that it can be taken more than once. In the meantime, we can only say that there does not seem to be any reason to suppose that the earth is special or that chemical life is not likely to have appeared many times over the history of the galaxy. Such life does not leave its planet unscathed and does not itself remain unchanged.[35] Suffice it to say that, once you have self-reproduction, then the entire mechanism of Darwinian selection and mutations takes over, and the evolution from a primeval slime mold to a local politician seems practically inevitable. We therefore expect that life will eventually become intelligent life in many cases.

But this expectation, in turn, raises another question. If there is this sort of inevitability in the progress from the hot, dense early universe to the formation of galaxies, stars, planets, and life, then there should be lots of other civilizations floating around space. Where are they? Of course, there are people who think they know the answer to this question—large silver ships drop out of the sky, land in their gardens, and take them for rides. These people generally have other problems as well. Others have tackled the question in a slightly different way. They start with the total number of stars in the galaxy, eliminate those that are too short-lived, too cool, or too close to another star to provide comfortable environments, and guesstimate factors for the probabilities of planets, the origin of life, and the development of civilization, by which they mean not the quartets of Mozart or the dialogues of Plato, but rather radio astronomy—that is, the ability to communicate across interstellar distances. Many such efforts end up with an estimate that a civilization may have appeared every $10^{2^{+2}_{-1}}$ years over the history of the galaxy.[36] A carbon-based biochemistry is assumed, but there is no presumption (and little likelihood!) that the products would look like us. Other kinds of biochemistry may possibly increase the numbers of civilizations above these estimates.

The question of whether we have millions of neighbors, thousands, or none then reduces to the question of how long a civilization lasts. Possible answers range from 10^2 years (the time over which we seem capable of advancing [?] from the first radio broadcasts to self-destruction) to 10^7 years (the time scale for biological evolution from one species to another among the higher mammals) on up to 10^{10} years (the Main Sequence lifetime of the host star). These answers correspond to our having no company at all, other civilizations within a thousand parsecs or so, or near neighbors within 100 or even 10 parsecs, corresponding to round-trip light travel times comparable with human lifetimes.

Some effort has been made to test this last, most optimistic, hypothesis by pointing large radio telescopes in the directions of nearby, solar-type stars and listening for a while.[37] No positive results have been reported. This is not particularly discouraging, because the sensitivity was such that nothing could have been seen unless a comparably powerful telescope, operating in a radar, broadcasting, mode, was pointed directly at us. As new radio facilities have come on line, the range of such experiments has gradually increased without yet finding anything, but also without yet reaching a level where we could see a planet like the earth in the absence of deliberate broadcasting. We could do this with present technology, for planets within a few hundred parsecs, though the cost would be comparable with the cost of the Apollo program or a very small war. This is, therefore, perhaps a good time to start thinking about the consequences for society (including zealous companionship in scientific research) of the possible discovery that ours is merely one of many civilizations, and (given the time scales involved, inevitably) rather a backward one at that. The sociological implications may be worth worrying about.

Universal Constants and Their Importance

Whether we are alone or one of thousands or millions of civilizations in the galaxy, we seem to be a fairly natural product of the total history of the universe. It is, therefore, of some interest to ask whether any universe would have done, or is ours special. This is quite different from asking whether the earth and sun are special, because there are lots of stars and one can do some kind of statistics. The universe, on the other hand, is by definition unique in our experience. One can, however, characterize the universe in terms of a fairly small number of properties, which lead to the history we have discussed, and ask what would have happened if one or more of these properties had been different from what we see it to be. We will require eight parameters—four small-scale ones and four large-scale ones (Table 2). The four small scale properties are the four forces of nature, the four ways in which one bit of matter or radiation can interact with another. Understanding the internal structure of the so-called elementary particles (protons, neutrons, mesons, etc.) may require a fifth force or even an entirely different way of looking at the problem,[38] but we cannot say anything very concrete about its properties or possible variations at the moment. These four forces are the gravitational, electromagnetic, nuclear (or strong), and weak interactions, in order of decreasing familiarity. Table 2 lists them in order from strongest to weakest.

Most of the dire consequences of changing the forces mentioned in Table 2 require changes of an order of magnitude or more and thus do not rule out a factor of two decrease in G over the past 10 billion years,[39] though this can probably be ruled out in other ways.[40] But changing the electromagnetic force (the charge on the electron) by even a factor of three would mean that water could not be a liquid at any temperature, while very small changes in the ratio of the electromagnetic to nuclear force would cause helium burning to produce either all carbon or all oxygen rather than a mix of the two.

Similarly, looking at the universe on the largest scale, we find four important parameters. Two of these are the numbers that told us which of the Friedmann models we live in. They are H_0 and ρ_0 and measure the age and average density of the universe. A third is the entropy of the universe, which can also be expressed as the ratio of the number density of photons to the number density of heavy particles (protons and neutrons). This defines the present temperature (which we measure with some precision) and, therefore, the temperature and density at which helium was formed in the big bang. Finally, we have the homogeneity and isotropy,

Table 2. Universal Forces and Properties and Consequences of Varying Their Values

Force	Phenomena controlled	Consequences of lowering	Consequences of raising
Nuclear or strong	Structures and reactions of atomic nuclei	Early universe converts all matter to heavy elements; no source of energy for stars.	No nuclear reactions at all or none past helium; no heavy elements made, so no chemistry.
Electromagnetic	Structure and interactions of atoms and molecules	Electrons not bound to atoms; no chemical reactions possible.	Electrons inside nuclei; no chemical reactions possible.
Weak	$v_e + n \leftrightarrow p + e^-$ and other beta decays	No hydrogen burning possible; no source of heat or heavy elements.	Early universe converts all matter to helium; no energy sources for stars.
Gravitational	Structure and dynamics of planets, stars, and galaxies	Stars don't get hot enough for nuclear reactions to occur.	Nuclear reactions so rapid that star lifetimes are very short.

Property of universe	Value	Consequences of lowering	Consequences of raising
Rate of expansion	$H_0 \sim 55$ km/sec/Mpc	Matter all comes out of early universe in dense configurations.	Galaxies can't form; matter ends up spread out uniformly.
Average density now	$\rho_0 = 10^{-32} - 10^{-30}$ g cm^{-3}	Galaxies can't form; atoms very lonely.	Early universe turns all matter into heavy elements; no energy sources.
Entropy or temperature	$\eta = n_\gamma/n_{baryon} \sim 10^{-9}$ or T = 2.7 K	Early universe turns all matter into heavy elements; no stellar energy source.	Galaxies can't form due to radiation pressure.
Isotropy and homogeneity	$\Delta T/T < 10^{-3}$	Galaxies can't form in anisotropic expansion.	

which we originally put in as one of our cosmological assumptions, but which are also observed to be approximately true and seem to be necessary for galaxy formation.[41] The changes in these properties required to produce the dire consequences are often several orders of magnitude, but the constraints are still nontrivial, given the very wide range of numbers involved. Efforts to avoid one problem by changing several of the constants at once generally produce some other problem. Thus we apparently live in a rather delicately balanced universe, from the point of view of hospitality to chemical life.

Implications

It seems, in other words, that the universe must be more or less the way it is just because we are here. This is one version of the anthropic principle (*"cogito ergo mundus talis est"*).[42] It is of some interest to try to understand what it might mean that our universe should fall within a rather narrow range favorable to life, for each of several parameters. It may just mean that God (or the Initial Conditions, depending on your point of view) has been very careful. This is a possible answer—it may even be the right one—but it is not a scientific answer, at least in the narrow sense that it does not lead one to make any further observations, do any further experiments, or carry out any further calculations.

There are, of course, other possibilities. It could be, for instance, that ours is merely one of many universes, and that it is only those very few with particularly favorable properties that ever develop living creatures who ask these strange questions. There are two senses in which this could be so. If our universe is, in fact (contrary to the preponderance of the evidence, but by no means impossibly), of the sort that will eventually turn around and recontract, one might imagine a (finite or infinite) series of cycles of expansion and contraction, each constituting a separate universe and each having its own values of the fundamental constants. This violates our initial assumption that general relativity is the right theory of gravity, because, within its framework, when you once achieve singular conditions (like infinite density) out of nonsingular ones during the first recontraction, you can never get out again.[43] Perhaps this should not worry us too much, though, because general relativity is a classical (nonquantum) theory, and at sufficiently high densities, quantum-mechanical effects will become important[44] and must profoundly change the nature of gravitation.[45]

The alternative, if our universe is an infinite, ever-expanding one, is that a number of four-dimensional space–times (universes) might be imbedded in five- (or higher) dimensional space, existing simultaneously, from the point of view of a five- (or higher) dimensional observer. There is no easy way to picture five- (or higher) dimensional space with a three- (or lower) dimensional mind, and it is probably hopeless to try.

A third possibility is that, at some time in the distant future, when we have understood "all" the physics of the universe, it will be obvious that the various quantities must have the relationships they do. There is some hint that this might be so in the "large numbers." These are several simple combinations of the constants in dimensionless form that, depending on just how you write them, are all of order unity, 10^{40} or 10^{80}. For instance, the radius of the observable universe ($c/H_0 \sim 10^{28}$ cm) divided by the classical electron radius ($e^2/m_e c^2 \sim 10^{-12}$ cm) is about 10^{40}. And the ratio of the electromagnetic to gravitational forces between an electron and a proton ($e^2/Gm_e m_p$) is 2×10^{39}, and the number of particles in the observable universe, N_0, is about 10^{80}, i.e. $(10^{40})^2$. Alternatively, these can be written as $8\pi G\rho_0/3H_0^2 \sim 1$ (this is equivalent to saying that the universe is not far from the boundary between ever-expanding and recontracting models) and $(c/H_0)/(e^2/m_e c^2) \sim N_0^{1/2} \sim 10^{40}$. The largest number of constants appears in $j(j + 1)\hbar c/e^2 = 1n\,(j(j + 1)\,\hbar c/Gm^2)$,[46] where, in all of these expressions, H_0 is the present value of Hubble's constant, m_e, e, and j are the rest mass, charge, and spin of the electron ($j = \frac{1}{2}$), G is the constant of gravity, c is the speed of light, ρ_0 is the present average density of the universe, $\hbar = h/2\pi$ and h is Planck's constant, and $m = m_e m_\mu/(m_e + m_\mu)$ where m_μ is the rest mass of the muon.

There are other numbers (involving stars and such) that come out $\sim N_0^{1/4}$ and $N_0^{3/4}$. Assorted interpretations of these large numbers have appeared from the time of Dirac and Eddington to the present.[47] Some of the numbers may only express conditions for the formation of stable stars with reasonable lifetimes. Others, especially $G\rho_0/H_0^2 \sim 1$ (which implies that G must change if it is to be true for all time) and its variants, have been made the cornerstone of whole new nonrelativistic cosmologies (typically not in very good accord with observations). Perhaps when we have learned enough physics, we will understand that these relationships must hold and why.

Finally, it may be that the complexity and nonpredictability that we call intelligence (in ourselves, and orneriness in our adolescent children) is an inevitable result of just having enough particles interacting. By way of analogy, Feynman[48] proposed considering a water molecule, whose structure, energy levels, and so forth can be calculated with some precision by the methods of quantum mechanics. But nothing in that calculation would ever lead us to predict waterfalls. The waterfall is a result of very many particles interacting in ways we cannot, in practice, predict or calculate. Similarly, perhaps, whenever there are enough particles interacting, no matter what laws or forces govern their behavior, a sort of complexity results that we would acknowledge as a fellow intelligence. One might, therefore, imagine a universe in which the early, hot state had turned everything into neutron star material, and the exceedingly compact creatures living there would contemplate a universe like ours and claim (in very low voices, presumably) that it couldn't possibly have any intelligent life in it because the density was too low.

Dark Matter

We are the stuff of which stars are made. Or are we? The question, seemingly settled more than a century ago, is once again open.

To the ancient Greeks, it was self-evident that heavenly objects consist of a fifth element, or quintessence, fundamentally different from the familiar earth, air, fire, and water. Fashions change, though. And in 1835, when a distinguished philosopher (Auguste Comte) said about the stars that "never by any means will we be able to study their chemical composition [or] mineralogic structure," he meant a chemical composition of iron, oxygen, carbon, and silicon, like the earth's. He was wrong, of course—not about the assumption, but about his main point. The means had existed, at least since 1815. That year, Josef von Fraunhofer recognized the great similarity between a pair of yellow (absorption) lines in the solar spectrum and the pair emitted by many laboratory flames. We now know it indicates the presence of sodium.

Science, then as now, was international. Thus by about 1862, Léon Foucault of France, Anders Ångström of Sweden, George Stokes and William Huggins of England, Angelo Secchi of Italy, and especially Gustav Kirchoff of Germany had all contributed to the idea that we can recognize individual chemical elements both on the earth and in the sun and stars by the wavelengths of light they emit and absorb. Heavenly objects were made of earthstuff and could be reliably studied via the light shining from them.

Other developments during the nineteenth century strengthened this conclusion. Tenuous gas between the stars was also emitting light from common elements. And the first apparent counter-examples turned out to fit in after all. These were the bright stars Sirius and Procyon, whose paths through space are not straight lines, but show tiny wiggles when watched for decades. The German mathematician Friedrich Wilhelm Bessel announced in 1844 that the wiggles were the effects of invisible companion stars, moving in mutual orbits with the bright ones. No one, however, supposed that these nonluminous objects might be made of "quintessence." In due course, the American telescope maker Alvan G. Clark ground a 12-inch lens able to separate off the light from the companion of Sirius, which he was the first to see in 1878.

Thus astronomers continued to expect to be able to see at least some light (which eventually came to include radio waves, x rays, and the rest) from all cosmic objects and to find them to consist of ordinary materials. This expectation was well in place when the first modern evidence for nonluminous matter surfaced in 1932. By way of explanation, nonluminous matter means the same thing as dark matter, and the words are used interchangeably. It is sometimes also called "missing mass," but this is a shade misleading. As we shall see, the mass is almost certainly there; it is the light that is missing.

The new evidence came from two rather different kinds of studies and was announced by two very different kinds of astronomers. One was Jan Oort of the Netherlands (who is, as I write this, still an active member of the astronomical community and recipient of many prestigious prizes). He had looked at the positions and motions of stars outside the disk of our own Milky Way galaxy and used them to calculate how much mass the disk must contain in order to produce those motions through the action of ordinary Newtonian gravity. This amount (about 0.1 solar masses per cubic parsec) is called the Oort limit, and its implications were taken seriously from 1932 onward. Then he added up the masses of the stars visibly present in the disk. The sum came out smaller than the amount required by about 50 percent, unless he added in a large correction for faint, uncounted stars. Many later studies have confirmed the discrepancy and, in addition, ruled out many forms of gas and the large numbers of faint stars that Oort invoked to make up the extra mass. Its nature remains a mystery.

The other astronomer was Fritz Zwicky, born in Bulgaria, holder of a Swiss passport, but by then working largely in the United States. He, too, had measured positions and velocities, but for galaxies in large clusters. This represents a scale 10,000 times or so larger than the one investigated by Oort. But the result was somewhat similar. The total mass required to explain the motions was larger than the sum of the individual galaxies in the clusters, this time by a factor of something like 100. Though other observers soon made similar measurements, Zwicky's discovery was not taken nearly as seriously as Oort's. Most astronomers from the 1940s through the 1960s suspected that the calculations deriving large masses from the measured motions were not appropriate to the clusters' real structure.

Sadly, Zwicky died in 1974, just about the time most astronomers began to appreciate the ubiquity and importance of nonluminous matter, and did not have a fair opportunity to say, "I told you so." (More happily, Zwicky had been able to say this in 1969 when a neutron star was found in the Crab Nebula, just as he had predicted it should be back in 1934.)

The recognition of dark matter in the universe had some aspects of what historians of science call a paradigm shift. That is, it involved not just new data but also a new way of looking at old data and problems that had been around a long time, especially questions about the formation, evolution, and structure of galaxies. The seminal event in this paradigm shift seems, in retrospect, to have been publication of a pair of short papers by trios of American and Soviet astronomers. Neither presented much new data. But each had assembled disparate bits of information, the collective import of which was that the mass associated with a typical galaxy extends to much larger distances from the center than does the light. Thus the total mass you measure for a galaxy depends on the distance from the center at which you are able to probe, and increases linearly with that distance out to 300,000 light years or thereabouts.

Today's database has most of the same kinds of files in it as the 1974 one, but, of course, many more items per file. Let's look into some of them, starting with the smallest scales and objects closest to us, and see what the data look like and the nature of the dark matter implied.

First, the discrepancy identified by Oort persists. The motions of nearby stars reflect the gravitational attraction of about 50 percent more mass in the galactic disc than we can find in the form of stars and gas. The inventory has become much more complete than it was in 1932. New techniques of radio, infrared, ultraviolet, and x-ray astronomy allow us to detect gas no matter what its temperature, and to count individual stars fainter than the sun by a factor of 100,000 or more. Neither gas nor very faint stars contribute more than about 10 percent of the Oort limit.

What else could there be nearby? Brown dwarfs, perhaps (spheres of gas just a little too small to burn hydrogen and shine like real stars; so far we have not seen even one of these). Or old, faint white dwarfs (remnants of a generation of moderately massive stars that died long before our sun was born). Or maybe black holes (but only tiny ones with masses like planets and diameters of a few centimeters, or enormous ones with masses up to a million times that of the sun and diameters of a million miles; stellar-mass black holes, or neutron stars, won't do because gas falling on them would make brighter x-rays than we see). Or, just possibly, one of the more exotic candidates (see accompanying table) for dark matter on larger scales.

Second, we can look at whole galaxies, spirals, like our Milky Way, ellipticals, and irregulars, like the Magellanic Clouds. The techniques are many. But the general idea always is to measure something which is moving around the galaxy at a large distance from its center and which responds to the gravitational force (hence the mass) of the central galaxy. For spirals, the definitive data come from rotation curves. These are plots of the orbit speed of gas in the disc of the galaxy as a function of distance from the center. The speeds are found from the Doppler effect—a small change in the wavelength of light we receive that is directly proportional to the speed of the gas toward or away from us. Once you know the orbit speed, V, and the distance from the center, R, the calculation is so simple I can't resist showing it. To within a small factor, the mass of the galaxy inside the orbit of the gas you measure, M, is just V^2R/G, where G is Newton's constant of gravity. Because most rotation curves show a constant value of V over a wide range of R, the mass found this way typically gets bigger out to the largest R where you can record anything. And the total often exceeds the summed masses of visible stars and gas by a factor of 5 to 15. In other words, there is more dark matter than luminous in whole galaxies. Gory details of gas and star velocities in spirals suggest very strongly that most of this dark matter is not in the bright, flat disk, but rather in the faint spherical halo traced by globular clusters.

Elliptical galaxies are a bit trickier. Most of them do not have enough gas to radiate bright, easily measured lines. And most of the big ones do not rotate. Thus we have to use absorption lines in the spectra of their stars to measure the Doppler effect. Instead of plotting a tidy rotation curve with a very simple calculation to follow, all we can plot is the range of stellar velocities, from which the calculation of mass is much more elaborate. Still, the result is much the same. The further out in the galaxy you look, the larger the mass you calculate, with the total nearly always much larger than the sum of the stars you see. Confirmation of the large masses comes from two subsets of ellipticals, those with companion galaxies or globular clusters in orbit around them (for which we can measure velocities and so masses by V^2R/G), and those with very hot gas in them. Such gas emits x-rays, from which we can figure out the gas temperature. The temperature determines the velocities, V, of individual gas atoms at distance R from the center of the galaxy and so again the mass.

Candidates for dark matter in whole galaxies cover the same wide range as those for the Oort limit. The exotic ones look more interesting, though, because their inclusion in models of galaxy formation helps to solve an old problem (remember this was one of the things a paradigm shift is supposed to do!). The problem is the very basic one of how galaxies form. The universe is very smooth on large scales. Radiation tracing ancient large structure varies less than 0.01 percent around the sky. Yet the smaller scales of galaxies and clusters are very lumpy (and a good thing, too, or we would not be here to worry about it all). The difficulty is to understand how the lumps could have grown to their present sizes, masses, and clustered arrangements without ruffling up the background radiation. Dark matter, especially dark matter occupying bigger volumes than visible stars in galaxies, helps precisely because it is not so concentrated and lumpy as the light, and thus not so likely to ruffle things. The exotic dark matter candidates help even more, because they do not scatter radiation very effectively. Quite big lumps can, therefore, grow in the dark material without disturbing the background. Then, later, ordinary gas flows into those lumps to produce the galaxies and stars we see. Dark-matter particles behave somewhat differently depending on whether they are fast or slow moving when galaxies form, hence the distinction between "hot" and "cold" candidates in the accompanying table.

The third file contains data for pairs and small groups of galaxies, including our own Local Group. As we move to these larger scales, both the measurements and the calculations become more difficult in practice, though much the same in principle. The observer records the positions of galaxies in the sky and the Doppler-effect velocities. The catch is that these are not exactly the R and V needed. Each is a projection—one or two dimensions of the three-dimensional quantity wanted. Correction for this is possible. But the amount depends on whether you think the orbit of a pair of galaxies is most likely to be a circle, a long, skinny ellipse, or something in between. A standard guess is a mix of these (called an isotropic velocity distribution) which yields total masses for binary and group galaxies exceeding the luminous masses by factors of 10 to 30.

For the Local Group, we can do a different, supporting calculation. The Milky Way and Andromeda galaxies are now moving toward each other at a speed of about 100 kilometers per second. Yet they must have started life moving apart, since the universe as a whole has been expanding for 10 or 20 billion years. Thus the present motion probably comes

Exotic Dark-Matter Candidates

Candidate particles	Approximate mass	Predicted by	Astrophysical effects
G depends on distance	——	non-Newtonian Gravitation	mimics DM on large scales
Λ (cosmological constant)	——	general relativity	mimics critical density
axion, majoron, goldstone boson	10^{-5} eV	QCD; symmetry breaking	cold DM
ordinary neutrino	10–100 eV	GUTs	hot DM
*light higgsino, photino, gravitino, axino, sneutrino	10–100 eV	SUSY/SUGR	hot DM
paraphoton	20–400 eV	modified QED	warm/hot DM
right-handed neutrino	500 eV	superweak interaction	warm DM
*gravitino, etc.	500 eV	SUSY/SUGR	warm DM
*photino, gravitino, axino, mirror particle, simpson neutrino	keV	SUSY/SUGR	warm/cold DM
*photino, sneutrino, higgsino, gluino, heavy neutrino	MeV	SUSY/SUGR	cold DM
shadow matter	MeV	SUSY/SUGR	hot/cold (like protons)
preon	20–200 TeV	composite models	cold DM
monopoles	10^{16} GeV	GUTs	cold DM
pyrgon, maximon, perry pole, newtorites, Schwarzschild	10^{19} GeV	higher-dimension theories	cold DM
sypersymmetric string	10^{19} GeV	SUSY/SUGR	cold DM
quark nuggets, nuclearites	10^{15} g	QCD, GUTs	cold DM
primordial black holes	10^{15-30} g	general relativity	cold DM
cosmic strings, domain walls	10^{8-10} $M_\odot$	GUT's	promote galaxy formation but contribute little density

*Of these various supersymmetric particles predicted by assorted versions of supersymmetric theories (SUSY) and supergravity (SUGR), only one, the lightest, can be stable and contribute to the density of the universe. The theories do not at present tell us which one it will be or even the approximate mass to be expected. Of the other abbreviations, QCD is quantum chromodynamics (the present "best buy" theory of the nuclear force); QED is quantum electrodynamics (the relativistic quantum theory of the electromagnetic force); and GUTs are grand unified theories (possible overall theories of the electromagnetic, weak, and nuclear forces).

from the masses of the two galaxies pulling on each other. We can combine the approach speed, the present separation (about 650,000 parsecs) and the time available for the motion to calculate the mass needed to do it—4 to 7×10^{12} solar masses, like that found for many other pairs and small groups. Dark matter is thus confirmed as exceeding luminous by 10 to 30 times on this scale. The same wide range of candidates remains possible.

Fourth, Zwicky's work on large clusters of galaxies is at last taken seriously. The Coma cluster he studied and about 100 others have had velocities and positions of their component galaxies measured and used to estimate masses. As in the binary case, the values we measure are projections of the ones we want, requiring a correction factor. In addition, the clusters are so big that the galaxies very far from each other may not yet have had time to respond to each other's gravitational forces. This requires another correction factor. You get it from a computer simulation, in which many masses, representing the galaxies or their precursors, are plopped down in space and allowed to interact with each other for 10 or 20 billion years. With luck, this is compressed into a few hours of computer time. As you can guess by now, the calculated total masses for the clusters exceed the luminous ones by still larger factors—100 to as much as 300. Even if you suppose that the mass properly belonging to a single galaxy is that given by a spiral rotation curve (which was already 80 or 90 percent dark matter), the clusters are still too massive by another factor of 10 or so, and must average about 99 percent dark matter.

Fifth and finally, some scientists are pretty sure that the total mass everywhere in the universe must be about 1000 times that of the luminous part. The corresponding density is only about one atom per bathtub volume, but it is just enough for the collective gravitational force to reach a standoff with the present universal expansion, so that the expansion continues for an arbitrarily long time, but at an arbitrarily small speed. A popular picture of the very early universe, called inflation, requires the average density to have very nearly this critical value. I am not myself entirely convinced that the inflationary scenario is the correct one, though it can be made to sound very persuasive, especially when explained quickly with a slight British accent.

At this point, we must begin to take the exotic candidates still more seriously. The argument comes from examining nuclear reactions that occurred 10 or 20 billion years ago (after inflation but before galaxy formation), when the universe was hot and dense. Under these conditions, you do not have ordinary chemical elements. Instead, there is a sort of soup of separate protons, neutrons, electrons, and radiation (not to mention positrons, neutrinos, and other oddities that need not concern us here). As the universe expands and cools from the hot state, the individual particles interact to form hydrogen, its heavy variant deuterium, two kinds of helium, and just a bit of the light metal lithium.

How much of each you get depends on how many protons and such were there to interact in each volume of space. For instance, the more particles there are, the easier it is for them to find each other. Thus higher densities tend to make more helium, at the expense of hydrogen and, especially, deuterium. If, then, we knew just how much helium, deuterium, etc. had been made, we would know the amount of proton–neutron soup and so the average density of the universe in the form of ordinary matter. Unfortunately, no pure early-universe gas remains around to be measured. It has all been polluted by additional nuclear reactions in stars (another thing to be grateful for, since we are made of carbon, nitrogen, and other pollutants!). But, looking at the purest stuff left and subtracting off the stellar contributions, we find rather more deuterium (about 2 atoms for every 100,000 atoms of normal hydrogen) and less helium (somewhat less than 25 percent) than one might have expected.

The amount of normal matter needed to make these abundances come out right is between 50 and 150 times the amount of luminous mass. Apparently, then, some of the dark matter must be ordinary stuff in the form of brown dwarfs, white dwarfs, or whatever. The clusters and superclusters of galaxies are on the ragged edge, needing dark/luminous matter = 100 to 300, while we are allowed about 150. And it would be very difficult indeed to accommodate a critical density, dark/luminous = 1000, in the kind of matter that participated in the nuclear reactions. There may yet be a way around the limit, if the universe was already quite lumpy (especially in its ratio of protons to neutrons) when the nuclear reactions took place. Several groups of astronomers, including Craig Hogan, James Applegate, and Philip Scherrer in the United States and Martin Rees in England, are currently exploring this idea, called inhomogeneous nucleosynthesis. Unless it proves very effective in encouraging deuterium and suppressing helium and lithium, however, we will still be forced to say that you cannot have a critical density of ordinary matter. As a result, anyone who thinks the average density of the universe really is the critical value must look long and hard at the alternatives to ordinary matter.

What can we say about the exotic candidates? Remarkably little from an experimental or observational point of view. Several different kinds are predicted by different branches of modern theoretical physics. Their names are many and varied. The popular collective nouns are "inos" (many of the candidates are related to known particles in a way that tempts one to invent words like *photino* from *photon* and *gravitino* from *graviton*) and "WIMPs," for weakly interacting massive particles. The weakly interacting part is what makes them useful for dark matter. Nuclear reactions among protons and neutrons carry on almost as if they weren't there; even the smooth background radiation is largely blind to them. The "weakly interacting" part is also what makes them difficult to study. Most WIMPs, most of the time, go right through the earth and sun, let alone a laboratory detector.

Experiments at Stanford and London have, just possibly, each seen one monopole (the magnetic equivalent of single electrically charged particles). And groups in the USSR and the USA have reported evidence for neutrinos having nonzero rest mass (the existence of neutrinos is not at issue, but only whether their masses are zero, like photons, or merely very tiny). Other rather similar experiments have not reproduced this evidence, and neither monopoles nor massive neutrinos are generally accepted at the moment. A dozen or more additional experiments are in progress or planned. Meanwhile, we can set some additional limits from the fact that WIMPs in large concentrations in stellar cores or galactic haloes will interact a little bit to produce light, hasten nuclear reactions, and so forth. These effects must not be too large, or we would already know about them. Thus, for instance, in different ways, the number of gamma rays radiated from the galactic center and the numbers of evolved stars in globular clusters rule out certain combinations of WIMP masses and interaction rates. None of these limits is currently tight enough to eliminate even one of the tabulated candidates.

Well, then, is the universe really made of earthstuff or not? Despite Auguste Comte, we know with great confidence that the luminous stars, gas, and galaxies consist of standard chemical elements—hydrogen, helium, and all the rest up at least to uranium—in standard proportions. But the universe as a whole is not very luminous. Somewhere between 90 and 99.9 percent of the mass does not radiate, and so its "chemical composition and mineralogic structure" cannot be studied directly.

If the dark matter outweighs the luminous by a factor of only(?!) 10 to 100, then it may well all be normal material, hiding in brown or white dwarfs or black holes. We cannot prove that it is, but have no strong reason to conclude that it isn't (apart from the difficulty of understanding galaxy formation). If, on the other hand, the inflationary scenario, or some other argument for a critical density, proves correct, then WIMPs or inos become serious contenders for the dominant constituent of the universe. Deuterium and helium come out right, and galaxy formation looks more straightforward (though, in honesty, one must say that the problems have not all been solved even by models that invoke hot or cold dark matter or a mix of the two).

How am I betting? At least 50–50 on protons and neutrons all the way. The remaining probability is distributed very thinly (though not quite uniformly) among the many entities in the table. Clearly one definitive laboratory experiment, recording the presence of axions, higgsinos, or whatever, could change the odds drastically, though the chances of any one experiment being much more definitive than the currents hints of monopoles and massive neutrinos are rather small. We are made of starstuff, in a narrow sense of the word, but whether or not most of the universe is made of star, earth, and people stuff is definitely once again an open question.

The Anthropic Principle

T he world is very full of things. Not just cabbages and kings, but stars and galaxies; fire and water; molecules, muons, and magnetic fields. The array quickly becomes bewildering. Thus, from the earliest times of which we have any written record, mankind has tried to make sense of it all by asking: which are the really important things, how do they relate to each other, what are their causes and purposes, and can we somehow control or at least predict them?

One approach to these questions is called religion (when it is practiced by our tribe) or magic (when practiced by the tribe in the next valley). It attributes causes and purposes to a creator or creators (somehow external to the world) and provides an organizing idea that the important things are those that are necessary for our salvation. This way of looking at the world can be made perfectly self-consistent and exceedingly resistant to disproof in the realms of organization, causes, and purposes. It does much less well in the realm of prediction and control. This nearly exhausts my knowledge of the subject, and I shall not further address it here.

Modern science is a very different approach to the same set of questions. The difference shows immediately in our use of the word universe rather than world. The universe is, by definition, everything there is—no external causes or forces need apply. The causes of objects, processes, and phenomena are understood, instead, as being other objects, processes, and phenomena, both equally part of the universe. And no purpose is sought or expected.

The enormous successes of science in the realm of prediction, explanation, and control require no inventorying. Reduction in tobacco use saves lives in a way that a ban on trumpet sales could not have saved the walls of Jericho. On the other hand, science provides no firm rule for picking out important things from the rich array. In fact, we run the risk of a sort of circularity, in which we regard as important precisely those items we can explain and neglect the ones we can't.

The anthropic cosmological principle provides a promising selection rule: the important events and objects are those that, if they had been very different from what they are, we would not be here to ask questions about. In addition, the idea that the universe must have certain properties in order for observers to appear can function as a sort of explanation or cause for phenomena that might otherwise seem to be happy accidents— at least until we can find a better explanation in terms of other phenomena. In both these cases, the anthropic principle serves as a sort of tool, enabling us to get on with the business of doing science.

How a tool works is often understood most clearly by watching it in action. Let us, therefore, sketch out part of the history of the universe and look at some of the events that the anthropic principle says are important, in the sense of being essential to the appearance of question-askers. I will focus on events and processes within the regime of astronomy, cosmology, and astrophysics, where my own scientific interests lie.

Our view backward extends 10 or 20 billion years, to a time when the universe was exceedingly hot and dense. The time scale is measured by the ages of the oldest stars, the decays of radioactive atoms (like uranium and thorium), and the rate at which clusters of galaxies are moving away from each other. The high temperature and density announce themselves via the products of nuclear reactions that occurred then and in the presence today of ubiquitous, left-over microwave radiation. Its current temperature is only 3 K ($-454°$F), but this implies billions of degrees in the past. Matter emerges from such hot, dense conditions in the form of nearly pure hydrogen and helium gases in ratio of 3 to 1. And it is lumpy. The energy emerges as light (electromagnetic radiation, including x rays, radio waves, and all the rest) and neutrinos (particles so disinclined to interact with anything else that their subsequent influence is negligible).

Which of these aspects of the early universe are vital? First, the time scale must be long enough for complex entities to arise. This puts limits on both the expansion rate and the density of the universe. Further limits on these and on the forces that cause nuclear reactions come from the need to have quite a lot of hydrogen among the products to serve later as fuel in stars. A cooler, denser, slower universe with stronger nuclear forces would have turned most of the matter to iron. In it, no stars would ever shine to illuminate and warm their planets. Too high a temperature would be equally fatal, for radiation would then prevent condensation of the lumps to make stars and galaxies. And these lumps, also, are absolutely essential. The average density of the universe today is about one atom per cubic meter. Thus, if the matter were spread out uniformly, you would be a poor, lone, lorn hydrogen atom; and in all your life you would never meet even one other hydrogen atom.

As the universe expands and cools from its hot, dense state, the most important process is gradual condensation of gas into stars. Stars perform two vital functions. First, relatively low-mass ones like our sun fuse hydrogen fuel to helium in their cores continuously over billions of years. As a result, the stars shine and, by shining, provide a long-term source of high-grade energy and the possibility of stable warm environments on planets orbiting them.

Second, stars more massive than our sun live shorter but more exciting lives, during which their cores get hot enough for additional nuclear reactions to produce carbon, oxygen, nitrogen, iron, calcium, and the other elements needed for the complex chemistry we call life. The first stars must have been nearly pure hydrogen and helium. Their planets, if any, would have been gas giants like Jupiter, and no chemical reactions beyond the formation and destruction of hydrogen molecules could have occurred on them.

The need for stars to shine down through the ages tells us that the force of gravity must be strong enough to hold them together against the outward pressure of their hot gases (pressure is one aspect of the force we call electromagnetism) but not so strong that the stars overheat and use up their fuel too quickly. Further limits on gravity, electromagnetism, and the nuclear forces come from the requirement that the elements cooked in stars must be blown back out somehow into the remaining gas. Then later generations of stars are formed from material that is no longer pure hydrogen and helium, and they can have planets like the earth, made of rocks, metals, water, and air.

Next, chemical reactions among atoms must be possible. This tells us something about the ratio between the strong force that holds atomic nuclei together and the electromagnetic one that keeps electrons in orbit around the nuclei. A factor of ten change in this ratio would prevent molecules from forming. A much smaller change would keep water from ever being a liquid at any temperature. There are, of course, other possible solvents for biological processes, but water has particularly nice properties (especially on a hot day).

Equally important later stages are the evolution of planetary atmospheres and the origin of molecules capable of self-reproduction, modification, and information storage (like DNA). Once these happen, the full mechanism of mutation and natural selection takes over, and evolution from a slime mold to a politician seems not implausible. These later stages lead also to interesting limits on the properties of a habitable universe. The details belong to the fiefdoms of geology and biology, and I am not competent to discuss them.

Thus the major stages in the history of our universe—from the hot, dense early phase, through galaxies and stars, to planets, chemistry, life, and a propensity for question-asking—all lead to constraints on what nature must be like. Most of these limits can be made quite quantitative, leading to firm numbers and equations. For instance, the ratio of strengths of electromagnetism and gravity is about 10^{39}; the ratio of the age of the universe to the time it takes light to cross an atomic nucleus is about the same; and their product (or the square of either) is roughly the number of particles in the observable universe. And if these things were not so, we would not be here to ask about them.

Thus the anthropic principle has functioned as a guide to identifying certain properties as important. But, in addition, it can be regarded as an explanation for them. Universes with other sets of properties may be conceivable, but no observers will ever arise to study them. This point of view is called the weak anthropic principle. It is, in a sense, merely a very extreme form of observational selection effect. You can only measure those things which you are equipped to measure. This is hard to deny!

A variant, called the strong anthropic principle, holds that observers must arise in any conceivable, real universe. If so, then a universe very much like ours is the only possible one, and the ratios I mentioned must be about 10^{39}. I find this form of the principle much less persuasive.

To recapitulate, the anthropic principle functions as a useful scientific tool in pointing out to us which phenomena are sufficiently fundamental that our mental picture of how the universe works (called a model by scientists in many disciplines) must account for them. The principle can also serve as a sort of explanation for these phenomena. But I am inclined to think it should be regarded, in this function, only as a sort of stop-gap.

Some examples may help to clarify this point. First, the number of atoms in a typical star is the three-halves power of the number, 10^{39}, that turned up when we looked at the age of the universe and the forces of gravity and electromagnetism. The stars would not shine usefully otherwise. This sounds like a new, interesting bit of information. But, in fact, it arises simply because the structure of stars is determined by the balance between these two forces; the number can be calculated from well-established equations.

Second, I mentioned earlier that the temperature of the universe must fall within rather narrow limits to permit both galaxy formation and the existence of hydrogen fuel for stars. These limits can also be expressed as a constraint on the ratio of the number of photons (light particles) to the number of baryons (matter particles) in the universe. The ratio is, and must be, close to one billion. This had, for some time, its only explanation in terms of something like the anthropic principle. But, in the mid-1970s, there arose mathematical ways of linking electromagnetism and the two (weak and strong) nuclear forces together, called grand unified theories, or GUTs. Each such theory predicts a fairly definite photon-to-baryon ratio, some of them equal to the observed, necessary value. The simplest of these theories can already be ruled out because it predicts something else that we don't see—decay of the protons in ordinary matter on a time scale of "only" 10^{29} years; we can set experimental upper limits near 10^{31} years, even though this is much longer than the present age of the universe. It is likely, however, that future more elaborate versions of GUTs will turn both the photon-to-baryon ratio and the proton lifetime into calculable, hence explicable, quantities. Then the anthropic principle will no longer be needed to account for them.

Looking still further ahead in time, we can foresee the development of more complete mathematical linkages among the forces, called superunification, supersymmetry, and supergravity. Within these, the strength of gravity and the age of the universe will probably also be calculable quantities. You might, of course, then say that the mathematical structures must be as they are or we could not be here to do the calculations. But this shoves the regime of the anthropic principle as an explanatory tool back nearly to the realm of "first causes," perilously close to the territory of religion, upon which I promised not to trespass further.

It's a Nice Planet to Visit, but I Wouldn't Want to Live There

As all dedicated vacationers know, you can stand anything for two weeks. But most of us become more particular when choosing a long-term residence, and downright fussy as we begin to think about where we want to raise our children. The macrocosm of life is rather like this microcosmic example. Terrestrial life in one form or another can survive for at least a while under a much wider range of conditions than those under which it could have evolved. And (we hope that!) human civilization will be able to endure circumstances that would surely have kept our ancestors cowering in their caves, or even wiped them out completely.

What, then, must a planet supply for life to be able to develop, survive, and evolve to intelligence on it? Charles Darwin's warm little pond is a good start:

> . . . conceive in some warm little pond, with all sorts of ammonia and phosphoric salts, light, heat, electricity, etc., present, that a protein compound was chemically formed ready to undergo still more complex changes . . .

That is, we seem to need liquid water (constraining both temperature and composition), a high-grade energy source, suitable raw materials (including carbon and hydrogen compounds, but no free oxygen), and lots of time.

Astronomers, looking at stars and the process of star formation from interstellar gas (which is surprisingly rich in a wide variety of carbon compounds), have generally concluded that these essentials should be available in many places; while biochemists have verified with laboratory experiments that the initial stages of the chemistry leading to proteins, nucleic acid, and so forth do actually occur in such a pond. The implications of these observations and experiments for the frequency of life-bearing planets in the galaxy have been frequently, and usually optimistically, explored by many people. It is usually assumed that time is not really a problem, because the oldest stars in our galaxy are three or four times the age of our sun. The catch is that these stars (hence, perhaps, their planets, if any) have only about one percent as much of the chemical goodies needed for life (carbon, nitrogen, phosphorous, etc.) as the sun does. Thus, the oldest chemically suitable stars may not be much older than our 4.6 billion year sun. A closer look at how life probably got started on earth suggests a number of other plausible constraints.

I want here to focus particularly on the extent to which the appearance of intelligent, technological life on earth may have depended on the existence of plate tectonics.

Plate-tectonics-and-continental-drift, which ascended from the realm of crankiness to that of dogma with extraordinary rapidity about 1965, is a sort of catch-all phrase to describe what most geophysicists now think are the dominant processes happening on the surface of the earth. Of the ideas that make up the dogma, the ones that will concern us here are:

1. Continental rocks (granites) are less dense than ocean basin rocks (basalts) and float upon them.

2. The earth's surface is broken into a number (about ten large and a few dozen small) plates of solid rock, typically containing both land and ocean basin, a few miles thick and hundreds to thousands of miles across.

3. These plates are dragged around on the surface of the earth essentially by the flow of less rigid (but still solid) rocks underneath them in the earth's mantle.

4. Where plates spread apart, molten rock rises and forms new ocean basins (this is happening now in the mid-Atlantic and elsewhere).

5. Where plates slide past each other, general unpleasantness occurs (the San Andreas fault of Southern California is the classic example).

6. Where plates crash into one another, continental rock is piled up into mountains (the Andes and Himalayas are regions where this is happening now; the Appalachians and Urals are relics of ancient collisions), while ocean basin rock can be dragged back down into the mantle and remelted, giving rise to volcanoes and considerable chemical reprocessing of the rock (as is now occurring around most of the rim of the Pacific Ocean).

7. This sort of thing has been going on for most of the age of the earth, breaking apart old continents, forming new ones out of the fragments, and generally messing things up thoroughly.

8. Energy to keep this going comes from the interior heat of the earth, which in turn is partly left over from its formation and partly maintained by the decay of radioactive atoms inside. Boundaries of the present plates are marked by the presence of volcanoes and earthquakes.

Suppose plate tectonic processes were to stop (as must eventually happen when the earth cools off) or had never occurred (as would be the case if the earth had started with a smaller mass, like the moon, or had contained smaller amounts of radioactive elements). Could we continue to live here, could we have evolved to something like our present biological state, and could our civilization (such as it is) ever have developed? The answer to all three questions may well be no. Let's look at three products of plate tectonics that seem to matter. These are (a) the maintenance of the balance between land and water, (b) the provision of a wide range of habitats, and (c) the production of high-grade metallic ores.

If plate tectonics turned off, the first to suffer would be the skiers and climbers, as the taller mountains eroded away over 10 million years or so. After a few hundred million years (this is still short compared to the 4.6 billion year age of the earth), essentially all the land would be eroded below sea level, and eventually the oceans should cover the entire earth's surface to a uniform depth of a few miles (as would always have been the case if there had never been any tectonics). One can imagine a portion of humanity developing under-water cities and surviving, but it is much harder to conceive of civilization arising under these circumstances. Dolphins, though endowed with impressively large brains in relationship to their body weights, are not tool makers or even (like many primates and other, lower, land animals) tool users. In fact evolution could not have taken anything like its actual course. Though many of our remote ancestors were ocean dwellers, they preferentially inhabited the shallow continental-shelf regions, not the ocean deeps, as do the vast majority of marine organisms to this day. Even the chemical reactions that were, in some sense, our earliest ancestors probably require shallow water, with sunlight and muddy bottom not too far apart. Perhaps we should stop here and conclude that plate tectonics is surely essential for life to develop. There may be a way out, though, via a planet on which water is less abundant than on earth (but more so than on Mars today), so that the surface irregularities produced by meteorite impacts are sufficient to keep some areas dry and others wet. Photographs of Mars suggest that it may have been like this not too long ago.

Plate tectonics introduces continuous change into both the number and types of habitats available for living creatures as continents and their shelves emerge, collide, and break apart again. Evolution can surely occur without this, but it would be different. Isolated populations, for instance on tropical islands, tend to be rather fragile, both flora and fauna being likely to succumb before species introduced from a larger, continental environment. In addition the fossil record shows interesting correlations with the positions of the continental plates in the past. When most of the land was in a single continent, diversity of species (both land and shelf dwelling) tended to be small, while existence of several smaller continents is accompanied by a wider range of living creatures. Evidently, when continents collide, species previously free of competition meet, and not all of them survive, producing a wave of extinctions (several of which appear in the fossil record over the past 600 million years). And when continents break apart, new, unoccupied habitats open up, and new species develop through adaptive radiation into the new territory. We are ourselves somewhat the product of such radiation, occurring when the opening of the Atlantic 60 million or so years ago separated the ancestors of the present New World monkeys from those of the Old World monkeys and apes. The result must in general be more rapid evolution than would otherwise occur. Given enough time, this shouldn't matter to the eventual appearance of intelligent life. But, as noted above, recent evidence suggests that there may not be very many stars that are both generously endowed with the elements needed for life and much older than the sun. And it has, after all, taken us the entire age of the solar system to get to where we are now.

Finally, most theories of planet formation suggest that the earth was originally chemically uniform, the iron–nickel core separating from the rockier outer layers as radioactive decay heated and melted the interior. This still leaves the rocky regions more or less well mixed. The wide range of minerals and ores we find on earth are, to a considerable extent, a direct result of melting, recrystallizing, dissolving and precipitating from water solutions of materials at and near plate boundaries (especially converging ones). Thus nearly pure copper occurs on Cyprus where the African plate is pushing into the European one, and lead, silver, and molybdenum ores are found in sequence across the western United

States, where the East Pacific divergent boundary has disappeared under North America. The development of some aspects of technological civilization would seem to be rather heavily dependent on availability of such ores. The example that comes immediately to mind is the need for good conductors of electricity if one is ever to discover Maxwell's equations of electromagnetism and the technology of radio broadcasting and receiving (the method most often suggested for interstellar communication). There are non-metallic conductors (including salt water and ourselves), but if the typewriter on which this is being written had to be plugged into the salt-water line, I think I would just as soon skip the whole thing.

Examination of other members of the solar system tells us that tectonic processes are by no means universal. The moon and Mercury are dead in the tectonic sense, their surface features largely attributable to meteorite impact. Venus and Mars show what may be dead volcanoes and faults but seem to have little or no activity now. The giant gas planets have no solid surface to be broken into plates. But the earth is not quite unique. In spring 1979, *Voyager* began sending back pictures of Jupiter and its moons. And one of these, Io, had no less than seven volcanoes in eruption at once (the terrestrial average is more like one). The Io volcanoes are not much like earth volcanoes, because Io has long since boiled away its water and similar light compounds (it is much less massive than the earth, so its gravity holds on to things less tightly). But they are volcanoes, suggesting the occurrence of chemical reprocessing (also indicated by apparent sulfur deposits on the surface), motion of surface rock, and so forth, as on the earth. The energy source is different, too. Io is not massive enough for left-over formation heat and radioactivity to keep its interior warm and fluid. Instead, it is continuously heated by friction of its rocks, which in turn is caused by Jupiter's very strong gravity pushing and pulling tidally on Io as it orbits the planet. The effect is rather like the heating of tires as they get compressed out of round and rub on the road driving at high speed. Apparently, then, any internal heat source will suffice to keep a planet's (or satellite's) surface active, and plate tectonics need not be confined to planets very much like the earth.

Thus I conclude tentatively that, although we might be willing to visit a planet whose interior is not warm enough to drive tectonic processing, we would probably not want to live there, and almost certainly could not have evolved there, but that such processing may be rather more common than one might at first have guessed. The reader may well want to draw different conclusions, and the foregoing words are meant only as a spur to further thought about the subject, and not as a final answer. Or even a final question.

For further information on earth processes see, for example, *Continents in Motion* by Walter Sullivan (second edition, New York: American Institute of Physics, 1991).

Where Are They?

"They shall mount up like eagles," dark-throated
assumes, Cold-sunned, low thunder and gentleness of
the authentic Throne.
VERNON WATKINS (1906–67), *Returning to Goleufryn*

B ut evidently not to earth, or such is the underlying assumption of many who believe that our failure to observe extraterrestrials has significant implications.

Less poetically stated, the argument goes that if intelligent life had evolved anywhere else in our galaxy, we would already have found out about it, in the same way the Indians of the Americas found out about Spain and Britain: by being colonized. We haven't been colonized (we think) or even received less unambiguous evidence of "their" existence; hence there is no intelligent life elsewhere in the galaxy.

The argument so stated clearly has gaps as wide as the Cumberland. Is the current absence of evidence for extraterrestrials really evidence of their absence? Let's look first at various aspects of that question and then go on to the implications of absence, like why isn't there any other life, and what should we do about it?

Have We Really Failed to Observe Extraterrestrials?

We have almost certainly not been colonized. All terrestrial living creatures, plants, animals, and one-celled creepy-crawlies share a common biochemistry and (very nearly) a common genetic code, strongly suggesting that we are all descended from a single origin of life. We are not half-native, half-colonist. In addition, as Cyril Ponnamperuma points out, that common biochemistry is very closely linked to conditions in the primitive earth oceans and atmosphere and to the reactions that could occur there, strongly suggesting that our origin of life was strictly homegrown. We are all natives of the earth.

Nor is colonization currently in progress. UFO sightings drift in and out of the popular press but are apparently not relevant to the question at hand. Robert Shaeffer has discussed some of the most widely publicized sightings. They are systematically explicable in terms of IFOs (identified flying objects, like balloons and satellites) or even NFOs (nonflying objects, like mushroom rings and street lights), with a fair-sized helping, in some cases, of human gullibility and craving for fame.

Less spectacular lines of inquiry haven't told us much so far, either. Ben Zuckerman, reporting on searches for electromagnetic radiation (mostly radio waves but also laser light and infrared radiation from Dyson spheres) from extraterrestrial civilizations, summarizes the situation: "Nobody has *reported* seeing anything." He should know, having been one of the two investigators (along with Pat Palmer of the University of Chicago) in one of the more extensive searches for patterned 21-centimeter radiation from nearby solar-type stars. But he has also remarked that, even if he and Palmer had seen something, they might not have told us about it, not being quite sure of the sociological implications of such an announcement. To put the searches so far carried out in perspective, it should be noted that they would fail by a factor of a million to detect the earth as a radio source from a nearby star and could only have seen anything if another civilization had been beaming a very powerful radio antenna directly at them at exactly the right time.

At this level, we have indeed failed to observe extraterrestrials.

If (Intelligent) Life Is Common, Should We Have Seen Something?

Michael Papagiannis suggests that space travelers would be so used to life in a capsule that they would never consider settling down on planetary surfaces when colonizing a new stellar system. Rather, they would build additional capsules out of local raw materials and continue to live in them. The most likely place for colonists in the solar system might, therefore, be the asteroid belt, and as a one-kilometer, self-contained space capsule would not look all that different from a one-kilometer-diameter asteroid, somebody may already be out there, without our recognizing them.

Why should there be colonists at all? The argument has two pieces, one mathematical and one philosophical. The mathematical part is addressed by Eric Jones and Michael Hart. If a civilization (feeling either population pressure or the call of adventure) sends out colonists who settle some place, flourish, and in due course send out more colonists, who in turn—and so forth—then it doesn't take very long to use up all the good planets in the galaxy, including, presumably, ours. The result follows largely from the rapid growth of any exponential function and does not depend very much on details of population growth rate, speed of space ships, or whatever, provided, of course, that space travel is possible.

Freeman Dyson and Cliff Singer discuss a range of interstellar propulsion systems. Some of these are considerably more elegant than the Project Orion method of throwing H-bombs out the back of the ship, for instance accelerating a spaceship by firing a stream of relativistic particles at it from ground and decelerating it by radiating shock waves into the interstellar medium. The important point is not that any one method is obviously the way to go. Rather, the fact that even our rudimentary technology can come up with a couple of possible ways of crossing interstellar distances strongly suggests that there must be other, better ways available to slightly more advanced civilizations. From a mathematical and technological point of view, then, we ought to have been colonized.

Sebastian von Hoerner and Ronald Bracewell address the philosophical point: Why should a civilization send out colonists in the first place? This sounds like it ought to be important. We can easily imagine (though not so easily achieve!) an alternative. A civilization might keep its population constant and balance its consumption of energy and raw materials with the rate at which these are supplied by its sun and planet (via processes in the earth generating new ores and so on). Another civilization may blow itself to bits early on, still others become fat and lazy or philosophically rather than technologically inclined. But, says Von Hoerner, none of this matters. If, as frequently advertised, a million or even a billion civilizations appear over the history of the galaxy, then the urge to colonize, even if very rare indeed, will surely strike one of them. And, concludes Bracewell, since diffusion is likely to be faster than evolution, whoever is out there first can probably preempt the entire galaxy. Perhaps, then, there really are not any other civilizations out there.

Why Are We Alone?

The standard scenario from the big bang to the birth of President Carter runs (1) hot, dense early universe; (2) lumps in the gas become galaxies; (3) stars form and make heavy elements (those besides hydrogen and helium, which come from the big bang); (4) second-generation stars that can have earthlike planets; (5) chemical evolution; (6) life; (7) biological evolution; (8) intelligence, technology, and Richard Nixon. Foul-ups might occur at any of these stages.

Stellar and Galactic Considerations

Terrestrial life has required a steady energy source, lots of heavy elements, and lots of time for its development. How common are these necessities? Because the abundance of heavy elements has increased only gradually during the life of the galaxy, it is not immediately obvious that many stars exist which can meet all the needs. There might be old stars deficient in heavy elements and young stars generously endowed with them, but few stars with both time and suitable raw materials for life-bearing planets. This would be a comforting answer, allowing us to be among the first civilizations on the scene for straightforward astronomical reasons. Unfortunately, it is apparently not the right answer.

Bruce Twarog and Pierre Demarque provide insights on the ages and composition of galactic stars. The average age of galactic disk stars is rather greater than that of our sun (6 versus 4.6 billion years), and most of those stars have heavy element abundances like the sun's to within a factor of two or three. Finally, stars of approximately solar mass, and therefore solar energy output, make up at least 10 percent of the galaxy. Suitable stars are evidently common.

Planetary Considerations

Earthlike life requires an earth-like environment. Reasonable minimum likeness might include a solid planetary surface with a protective atmosphere and liquid water (meaning temperatures between $0°$ and $100°C$ at least some of the time) persisting over a billion years or more.

How common are planetary systems? Frankly, we haven't a clue. Pat Harrington says that outside the solar system we have no evidence for any objects less massive than the smallest stars (about 0.06 solar masses or 60 Jupiter masses). Shiv Kumar suggests that, since as many as 90 percent of all stars may have binary stellar companions, planets might be very rare indeed. This may be correct, although orbit calculations (for example by Robert Harrington) show that planets can happily circle one or both members of many binary systems for billions of years. Whether or not they can *form* in binary systems is simply not known.

Jill Tarter predicts that we will be able to detect planets around other stars (if they exist) within ten years or so and, therefore, we need not argue about the issue. Unfortunately, even the most optimistic estimates say that we will be able to see only "Jupiters" and not "Earths" through their effects on the motion of their parent stars. Terrestrial planets may orbit half the stars in the galaxy or only one in a million. We cannot tell.

Even if terrestrial planets are common, livable ones may be rare. The two dangers are runaway ice ages, in which snow and ice reflect so much of the incoming starlight that they never warm up enough to melt again, and runaway greenhouses, in which everything evaporates and forms a blanket around the planet, keeping it too warm for water ever to condense and fall again as rain. The planet's life is made still more difficult by the fact that all sunlike stars get gradually brighter over the billions of years they live, and the planet must survive both the cooler early phases without icing up and the warmer late phases without total evaporation.

Papagiannis and Hart have both calculated the fraction of planetary systems in which at least one planet ought to be able to overcome all these hazards for at least a few billion years. Their answers are one in a million and one in 120, respectively. Our one bit of experimental evidence—that the earth has stayed much the same temperature (as judged from the fossil record) while our sun has brightened some 30 percent over the past few billion years—may suggest even a larger fraction than the latter calculation. Clearly more work is needed.

Given a sufficiently high level of technology, we can imagine transforming unpleasant planets into earthlike ones. The process, says James Oberg, is somewhat more complex than dumping algae into the atmosphere of Venus and waiting for them to turn the carbon dioxide into oxygen, but it can probably be done. This would affect the number of planet systems available for colonization, but not the number on which life could initially evolve.

The number of planets on which life can begin is presumably increased if biochemistry very different from our own is possible. Gerald Feinberg and Robert Shapiro offer some of the alternatives. Water, for instance, may not be the only or even best possible solvent (after all, it keeps trying to break up DNA). Liquid ammonia and liquid silicates could expand the livable temperature range to a $-50\,^{\circ}$C low and a $+1000\,^{\circ}$C high. At still lower temperatures (on planets orbiting red dwarf stars?), information might be stored in arrangements of the two forms of solid hydrogen. And these were two of the more conservative ideas!

In any case, planets, or habitable planets, could be much rarer than is usually supposed.

Biological Considerations

Darwinian evolution (self-reproduction with mutations and selection) is a marvelous way to build up extremely complex organisms with enormous capacity for storing and processing information. But how do you get started?

Hubert Yockey, Edward Argyle, and Hart calculate the probability of getting a reasonable protein or a reasonable strand of DNA, by assembling chains at random from a soup of amino acids or bases, in the lifetime of the earth. Any such calculation yields a very small number, less than one in a billion billion. Either life is extremely rare, not just in the galaxy but in the universe, or it wasn't done that way. Other ways have been discussed in the scientific literature. Strangely enough, our existence cannot be used to limit the probability of life very much. If we are unique (in a finite universe) or part of a sufficiently rare class (in an infinite universe) then probabilities like one in a billion billion per star or even smaller are not ruled out by our presence. My strong prejudice is that the origin of self-reproduction has a much higher probability and is achieved via a straighter path than random assembly of acids and bases. Laboratory biochemical experiments (of the kind called Urey atmosphere experiments, in which assorted raw materials like methane, formaldehyde, and ammonia are dissolved in water and fed energy for a while) may eventually indicate one or more possible paths. As Ponnamperuma remarks, he's only been working on the problem for twenty years, and nature had several hundred million; be patient.

Thus the probability of life per habitable planet may be anywhere from one down to the smallest number you care to imagine. In an infinite universe, there will be an infinite number of life forms, no matter how small the probability. But they will be very far away.

Richard Gott addresses the problem of communication under these difficult circumstances, suggesting that the best wavelength for broadcast and reception may be 5 to 6 millimeters, near the peak of the microwave background radiation, rather than the 21-centimeter wavelength of hydrogen. In any case, dialogue would not be possible among civilizations in different galaxies, as the roundtrip message time runs to millions or billions of years.

Finally, of course, it is possible that life, even intelligent life, may be common, but technology rare. One feels vaguely that the existence of whales and dolphins—intelligent but non-technological species on earth—must somehow be relevant, but it is far from clear how to use them to estimate the probability of technology occurring among extraterrestrial life forms.

Obviously, our failure to observe extraterrestrials is fully accounted for if life, intelligence, or technology are sufficiently rare.

What Should We Do About It All?

Many of the experts quoted here conclude that life (at least intelligent life) is rare in our galaxy. But this is partly a selection effect. Readers familiar with the literature of SETI will have missed the names of many of its best-known advocates among those whose views are discussed. Still, the non-life (or anti-life) point of view clearly has some justification and is entitled to have its consequences explored. As usual, there are two pieces: deciding whether it is right and deciding what to do about it.

The Future of SETI

Perhaps the most "anti" statement of all comes from Kumar, whose opposition is based partly on expense. A quick calculation shows that searches like the Palmer–Zuckerman one and that carried out at NASA-Ames by Jill Tarter cost $10 to $100 per star; if that is expensive, then you certainly shouldn't have bought this book. On any less extreme view, the only thing to do is keep looking—if we don't look, we are guaranteed not to find anything—and keep thinking about the problems between the Big Bang and interstellar travel.

The Future of Man in Space

Pogo once looked up at the sky and said, "Either there is somebody else out there, or there isn't. Either way, it's a sobering thought." It is indeed. If it turns out that human life is the most advanced form in the galaxy, then many of us would perceive it as even more valuable than we had previously thought and want to act accordingly. And if we are indeed alone, then we have the opportunity (the responsibility?) to, in Bracewell's words, preempt the galaxy and rebuild it as we see fit. Singer suggests that (if we don't blow ourselves up first) interstellar travel is possible in a hundred years, likely in a thousand, and unavoidable in ten thousand. But he also says that the most important thing is not when we get into space but the quality of life we send there.

LIVES OF STARS

Classifying Ourselves

Deciding just how normal or average our star and galaxy are is of practical as well as philosophical interest. The assumption that ours is a normal galaxy of a particular type is now being used in efforts[1] to pin down the extragalactic distance scale, the value of Hubble's constant, and the age of the universe, all still uncertain by about a factor of two. And properties of other stars are frequently determined by comparison with those of the sun. Thus, if spectral type, color, temperature, and luminosity of the sun are not known accurately, our ideas about how the chemical composition of the galaxy has evolved with time, about the mass–luminosity relationship, and even about the distance to the Hyades (which feeds back into the extragalactic distance scale) will come out wrong.[1-4] Two recent papers represent the latest round of squabbles over whether the Milky Way is an Sb or an Sc spiral and whether the sun is unusually red or normal for its spectral type.[5,6]

Hubble[7] and later workers[8,9] classified spiral galaxies into two sequences, with central bars present (SB) or absent (S or SA). Both sequences are ordered by the degree of prominence of the spiral arms, from Sa, with bright nuclei and inconspicuous tightly wound arms, through Sb to Sc, with very conspicuous, loosely wound arms. Many other properties of galaxies correlate well with Hubble type, including the zero points of the relationships (between total luminosity and rotational velocity, or surface brightness, or other distance-independent parameters) used to calibrate the extragalactic distance scale. The difference in luminosity between Sb and Sc galaxies at a given value of some other parameter is typically about a magnitude, corresponding to a 60-percent error in distance if you guess the wrong type. As a result, we can only calibrate these relationships using data from our galaxy if we know its Hubble type and that it is normal for its type with some precision. Unfortunately, we cannot photograph the Milky Way from outside and compare arms and nucleus optically. Thus all classifications of our galaxy are more or less indirect and correspondingly uncertain. The advent of radio astronomy greatly increased the number of indirect indicators available[10] and led Baade to decide that we live in an Sb galaxy.[11] Later determinations[5] have oscillated between Sb and Sc. The most detailed yielded the classification SAB(rs)bc,[12] indicating intermediate status in all parameters, including existence of a ring, r.

If we accept this intermediate type, then our galaxy is quite normal,[1, 12–14] having values of all the usual parameters intermediate among those of four other galaxies of similar types. A composite drawing of the four shows more than two spiral arms, not all of equal brightness. And Hubble's constant, H, must be near 100 km/sec^{-1} Mpc^{-1} or the coherent picture falls apart. This last conclusion disturbs many astronomers, as it requires either that the universe be less than 10^{10} years old or that Einstein's infamous cosmological constant be nonzero.

But suppose we are really an Sc. Galaxies of that type are intrinsically fainter than Sb galaxies. Thus when other galaxies are matched to ours, the effect is to make them intrinsically brighter, hence farther away. According to Paul Hodge,[5] an equally coherent picture can then be assembled with H near 50 km/sec^{-1} Mpc^{-1}, at least for luminosity and type determinations based on the scale length of the ionized hydrogen distribution in galactic disks. Classifications as Sc not depending on the distance scale[15–17] then become arguments in favor of the lower value of H. It remains to be seen whether other distance-dependent indicators of Hubble type can also make the Milky Way look like a normal Sc with the larger distance scale. De Vaucouleurs believes that this will not prove possible.

By contrast, the spectral type of our sun cannot be debated. Modern systems of classification define G2V stars as those whose spectra are like the sun's, at least at relatively low dispersion. During the 20 years from Kuiper[18] to Stebbins and Kron,[19] there seems to have been general agreement that many other G2V stars indeed had absorption line spectra like the sun's and were, on the average, the same color, $(B–V) = 0.63$, and so at the same temperature.

Metal abundances for solar-type stars in the Hyades and other young clusters using these similarities turn out larger than solar by 20 to 50 percent, implying that our galaxy is becoming richer in heavy elements as generations of supernovae pour out synthesized nuclei. Most of us would like to believe all this.

Doubts began to gather only when modern photoelectric measurements of the solar color[20–22] and indirect determinations, employing comparisons of spectral features with those of other stars,[23] started yielding significantly redder colors of 0.65 to 0.69, typical of G5, but not G2, stars. Hardorp added to the difficulties by searching among the brighter stars for ones that would match the sun in both broad-band energy distribution and detailed absorption line profiles.[2, 3, 6, 24, 26] Such solar analogues are rare[26]—fewer than a dozen out of several thousand stars searched—and nearly all are rather red, with $(B-V) = 0.65$ to 0.71. Most had been classified near G5, not G2; and three of the closest analogues are in the Hyades, implying that the sun and these stars are rather similar in their abundances of the heavy elements.

What can have gone wrong? Chmielewski[27] has looked again at indirect measures of solar $(B-V)$, obtaining a best value of 0.633 ± 0.25. This would permit the conventional wisdom to persist. He questions the accuracy of the direct measurements, all of which require some intermediary to bring small amounts of sunlight into a telescope and photometer that can also look at stars. But the most detailed direct determination yet[28] persists in finding 0.686 ± 0.01, using three independent intermediaries.

Could spectral classification methods have unexpected bugs?[3, 29] Hardorp has noted that the solar spectrum used as a standard is often that of sunlight reflected from Uranus or Neptune. Processes in the planetary atmospheres can partially fill in absorption lines, making the spectrum mimic that of a hotter star. Thus, when these artificially weakened lines are used as a standard, stars that really are like the sun are classified as cooler, near G5. Once this is allowed for, then several of the solar-analogue stars end up as G2 to G3,[6] and $(B-V) = 0.67 \pm 0.01$ becomes normal for the type.

But since the sun and its analogues in the Hyades then match in line profiles, detailed energy distributions, and broad-band colors, they must have the same temperatures and the same compositions. If the galactic metal abundance has not changed in nearly five billion years, then either supernovae have stopped expelling heavy elements (which sounds unlikely but is remarkably hard to disprove[30]) or the products must be diluted by gas from regions where no massive stars have formed (less unlikely, but also hard to prove).

In summary, to conclude that the Milky Way is a normal Sbc galaxy and that the sun is a normal G2V star, while philosophically satisfactory, leads to unexpected difficulties in understanding the scale of the universe and the evolution of our galaxy. This discussion may remind some readers of the revolutionary hypothesis advanced by Richard Armour that

Julius Caesar was written neither at the end of Shakespeare's first period nor at the beginning of his second period (an ancient topic of controversy) but rather between two periods.[31] If so, the blame attaches entirely to the present writer and not to those working on problems of stellar and galactic classification.

The Odd Two Percent

Tracking down the chemical evolution of galaxies, like any other scientific problem, has two pieces: first, figuring out what the world is really like and, second, trying to understand how it got to be that way. Thus we need to measure the amount of each element present in different kinds of stars, gas, and galaxies and to correlate these abundances with other properties of the objects, especially their ages and environments. Then we need to identify which nuclear reactions take place, when and where they occur, and how they and their products interact with other astronomical processes to produce the observed distributions. Given the magnitude of the task, the biggest surprise is perhaps that we have really made quite a lot of progress on it.

Cosmological Nucleosynthesis

The considerable disparity between amounts of hydrogen and helium on the one hand and of everything else (called heavy elements or metals) on the other reflects a vast difference in the conditions under which they form. Hydrogen and helium are left from a hot, dense stage through which all the matter in the universe passed 10 or 20 billion years ago. In contrast, sodium, chlorine, and all the rest are largely the products of nuclear reactions that take place only in the cores of rather rare, massive stars.

By way of reminder, the nucleus of an atom contains protons (the number determining which element you have, for instance 6 = carbon, 7 = nitrogen, 8 = oxygen) and neutrons (the number determining which isotope you have, for instance 6 or 7 in stable carbon, 8 in carbon–14, whose decay rate is handy for measuring ages of archaeological sites). Protons and neutrons are collectively called nucleons. Their production, nucleogenesis, occurred at a still hotter and denser phase of which we have relatively little knowledge. Nucleosynthesis is the array of processes by which they are assembled into nuclei.

The catch phrase "God made hydrogen and helium; Burbidge, Burbidge, Fowler, and Hoyle made all the rest" is a summary of those processes. Two graduate students, Ralph Alpher and Robert Herman, working with George Gamow in about 1950, first calculated the proportions of hydrogen and helium expected from cosmological nucleosynthesis in the hot early universe. They also predicted a radiation background much like that found in 1965 by Arno Penzias and Robert Wilson. Somewhat unfairly, modern astronomers generally cite only a brief publication, to which Hans Bethe's name was attached *in absentia,* to produce the euphonious combination Alpher, Bethe, and Gamow. The exact ratio of helium to hydrogen depends on how crowded the nucleons were when they interacted and so is one of our best guides to the total density of matter in the universe.

Stellar Nucleosynthesis

Gamow long suspected that the heavy elements had also been made in the early universe. We now know this doesn't work. But it was at least possible in a big-bang universe and quite impossible in a steady-state one. Thus the proponents of steady state were forced to think long and hard about alternatives (a classic example of how a wrong idea can prompt the doing of first-rate scientific research). The result was a detailed sequence of nuclear reactions that can occur in stars of different masses at different stages in their evolution. The massive 1956 paper by E. Margaret Burbidge, Geoffrey R. Burbidge, William A. Fowler, and Fred Hoyle (nearly always called B^2FH, all of whom still live and flourish) classified these reactions in essentially the modern form. Alastair G. W. Cameron did much the same thing at the same time, but had the bad luck to publish initially in classified documents (and to lack a memorable acronym).

The reaction sequence begins with the fusion (called burning, for no good reason) of hydrogen to helium in one of two related processes, both written down by Hans Bethe in 1939. All stars, including our sun, live on this for 90 percent or more of their lives. Next comes helium burning, whose dominant products are carbon and oxygen. This will eventually occur in all stars more than about half the mass of our sun (which is all of them that have had time to do anything interesting in the age of the universe). The vast majority of stars go no further, and trap the carbon and oxygen products in a dense remnant core, rather than sharing with the rest of the universe. Only the one percent or so of stars that start out with 6–8 or more times the mass of the sun go further. They complete hydrogen and helium burning in millions rather than billions of years. Then, as temperature and density at their centers continue to rise, carbon and oxygen fuse to produce neon, magnesium, silicon, and all the elements in between, and these, in turn, react to create elements on through to iron.

At this point, the massive star is in trouble. No further nuclear reactions can release any energy. Thus the iron core grows to a critical size and then collapses suddenly and catastrophically, releasing enormous amounts of gravitational energy. This energy goes into all sorts of useful things. A bit of it synthesizes many of the elements beyond iron (I'm rather fond of thulium myself, but you may prefer platinum or iodine). Some accelerates cosmic rays. Part comes out as light; and the first astronomer to see it sometimes gets his name in the paper as the discoverer of a supernova. Finally, some of the energy blows off the outer layers of the star. Thus the heavy elements that the star burned out its heart to produce are scattered to the corners of the interstellar medium, where they can be incorporated into later generations of stars and their planetary systems.

Our sun is a least a third-generation star. That is, it contains not only the dominant elements made in this basic scenario, but also some less common ones (like nitrogen and barium) that come from subsidiary reactions in stars that already have some carbon, iron, and so forth. From this remark, you can deduce that I have simplified what astronomers are pretty sure really happens almost beyond recognition. And at that it took several long paragraphs.

How do we know that this is really the right picture? First, only relatively young stars, fed by many generations of supernovae, have their full quota of 2% heavy elements. In the oldest stars that are still around, hydrogen and helium add up to 99.999 percent. Only a very few supernovae had contributed to the gas from which they formed. Second, looking in different environments, we see 2% or even 3% metals in initially dense regions, where star formation proceeded very quickly and was completed long ago. Low-density gas, on the other hand, forms stars much more slowly, and the heavy element abundance is only the odd 1 percent, or less, in such regions.

A Young Galaxy Ages

The combination of cosmological and stellar nucleosynthesis leads us to a model in which out of the hot dense early universe come three-quarters hydrogen, one-quarter helium, and lots of photons and neutrinos. The gas is necessarily lumpy. Otherwise, to this day you would be a poor, lone, lorn hydrogen atom with the next atom a foot or so away. The lumps gradually contract under their own gravity, twist each other into varying amounts of spin, and begin fragmenting into stars. This phase of galaxy formation is the current focus of an enormous amount of astronomical research, which means (contrary to what you might guess) that it is not very well understood. (An essential characteristic of a scientist is a certain contrary-mindedness that makes him willing to spend most of his time thinking about things he doesn't understand.) The hard part is to get enough lumpiness to make galaxies without disturbing the simultaneous smoothness of the left-over photon background. Many strange entities, from axions to decaying neutralinos to nontopological solitons, have been summoned from the depths to help with the problem. They have responded with, at best, moderate enthusiasm (a situation which may or may not remind you of a famous Shakespearean quotation).

In any case, the show was well underway by the time the universe had expanded to about one-fifth its present dimensions (or about 10 percent of its present age). Light from the centers of galaxies at that epoch reaches us carrying the signature of the usual heavy elements at perhaps a third of their eventual 2% strength. Not all galaxies age at the same rate. The ones that hosted these first quasars will, by now, be distinctly middle aged, all their gas long ago converted to stars. Corresponding nearby objects are the giant elliptical galaxies—massive, dense, metal-rich (3% or more, at least at their centers), and red (indicating both the atmospheric properties of metal-rich stars and the absence of young ones). At the other extreme are small, low-density galaxies, like the Large and Small Magellanic Clouds. More than half their visible material is still gas; formation of massive stars is vigorously underway; and their metal share is only the odd 1% or less. In between come spiral galaxies, like our Milky Way, consisting of several components. A dense, red, metal-rich nuclear bulge resembles the highly evolved giant ellipticals. A more tenuous halo has been drained of its gas, leaving only old stars of low metal abundance (there are also small whole galaxies, called dwarf ellipticals, in this condition). And the gas has settled down into a rotating disk, where star formation continues, and old and young stars, with a range of metal abundances, coexist. The disk thus resembles the Magellanic Clouds, but has more metals, owing to the input from earlier generations of halo stars. Another classic paper, published in 1962 by Olin Eggen, Donald Lynden-Bell, and Allan R. Sandage, quantified this relationship between heavy element abundance, location, and stellar motions (kinematics and dynamics, to be pompous about it) within our own galaxy.

Eating Bread and Jam

Wherein lies the difficulty that keeps large numbers of astronomers working away on issues of galactic evolution? It is the problem encountered by Winnie-the-Pooh in eating bread and jam (and by me, when I first started living away from home, preparing eggs, bacon, toast, and coffee for breakfast): How do you make everything come out even?[*] In the context of galactic evolution, this means that we must keep track of both the chemistry and the motions and locations of things at the same time, so as to get (1) the right total amount of heavy elements, (2) the right proportions of each (different ones coming dominantly from stars of different masses), (3) the right time history of each (since we can probe the past by looking at stars of different ages), and (4) the right proportions of nuclear bulge, halo, and disc, all in a galaxy of the right total mass, luminosity, and color, and all at a time = now.

Inevitably, the only possible approach is massive computer simulations. One's first thought is that with 200 billion stars to follow in a typical galaxy, the simulations will be massive even beyond the dreams of IBM and Cray. Finding a way around this barrier and so creating the first realistic models of galactic evolution was, far more than most major advances in astronomy, the work of one person. She was Beatrice M. Tinsley, and the work was her 1967 doctoral dissertation at the University of Texas, Austin. Her solution was to divide the stars up into about a dozen classes by mass and then to represent each class by a single track of brightness, color, and nucleosynthesis versus time.

Tinsley's initial models, for single, homogeneous zones, defined by a rate of star formation and a ratio of big to little stars, were a

---

[*]Not that it helps much with the astronomical problem, but Pooh's solution was to keep eating until the end of the chapter. When I asked my mother about the bacon-and-eggs version, she assured me that it would magically come out OK when I started cooking for a husband rather than some less regular arrangement. She was right.

startling success. They reproduced conditions not only in various present, nearby galactic environments but also in the past and distant ones that we must understand to learn about the large scale structure and evolution of the universe. By the time of her death in 1981, serious investigation of galactic evolution had grown from a one-woman cottage industry to a major branch of astrophysics, with practitioners in most large departments.

A typical model, 20 years after, allows for a number of additional effects. The most important of these are interactions between the zone you happen to be interested in and the rest of the universe. Gas (pristine or polluted) may flow in or get blown out. Encounters with other objects can cause bursts of star formation or (if too much gas is swept away) its total cessation. Stars with relatively long lifetimes can start feeding in special subsets of heavy elements long after you thought nucleosynthesis from that generation was complete. And (as the recent supernova 1987A has reminded us) the mass of a star is not the whole story. Its color and luminosity history and the mix of heavy elements it makes also depend on the composition it starts out with and can be changed by the presence of a companion star.

Looking Ahead

The main gain from these increasing complexities has been the ability to model a far wider range of the conditions that must have occurred at various places in galaxies. The main loss has been of uniqueness. That is, we can no longer look at the particular set of abundances in, e.g., the solar neighborhood and say that only one possible past history could have got us here. Different combinations of changing star formation rate, gas inflow and outflow, and proportions of high-mass and low-mass stars are all equally likely.

Where do we go from here? The answer may come from another person with the unusual insight and enthusiasm of Beatrice Tinsley. In the meantime, there is always hard work as an alternative. Most of the variables in the models are treated as adjustable parameters. That is, you guess several different values, run the simulations, and pick the answer you like best. But in a real galaxy, the formation of stars (both rate and range of masses) is a deterministic process, controlled by properties of gas—its temperature, density, composition, magnetic fields, and (probably) other things. Only we don't yet know enough to look at a particular gas cloud and say, "You are obviously going to form 1263 stars of the following masses in the next 100,000 years." Much the same thing is true of the other complexities.

Thus progress just now is coming from us harmless drudges who slog away on the details of small pieces of the whole picture. Do the metal-poor stars have even less than their fair share of nitrogen? Is there a correlation between the average metal abundance in a star cluster and the masses of the individual stars? Are the stars that form in close pairs a random sampling of the rest? (This is my own small corner.) Is there a separate, thick disk, component to the Milky Way, intermediate in properties between the halo and the star-forming disk? Does the . . . (At this point the editor slowly pulls down the curtain, blocking out the view of the author still babbling away questions to which she does not know the answer.)

Close Binary Stars

Most stars come in pairs. Wide doubles that can be separated in a small telescope are among the most familiar and charming of celestial sights. But many other binary stars are so close that they can never be resolved.

Close binaries might at first seem less interesting than separable doubles, but in many ways they are more so, since a great variety of dramatic and seemingly unrelated astronomical phenomena spring from them. The comfortable predictability of an Algol-type variable, the total unexpectedness of a nova, the strange chemical composition of an otherwise normal-looking star, the rapid flickering of an invisible x-ray source, and one kind of supernova explosion are among the many phenomena that can be produced by close binaries. In fact, many of these guises and others can be worn by a single pair at different stages in its life.

In a wide double—one whose stars are separated by at least a few astronomical units—each star lives out its life as if it were single. Its mass largely determines its color and luminosity, what nuclear fuel it burns and for how long, and the manner of its death. If the stars are close enough to interact, however, the pair evolves in ways more complex. The variety of stages the system goes through, and the strange effects produced during each, are often bewildering. But there is order beneath the confusion. To find it, we shall follow the evolution of close pairs that start out with various masses and separations. Along the way, most close-binary phenomena will fall into place.

The reader is warned that this discussion applies only to population I stars—those less than ten billion years old, of roughly solar composition, and born in the disk of our galaxy—and not to the older, metal-poor population II stars of the Milky Way's halo. There, binaries seem to be rare and to form, evolve, and die in other ways. Even among the population I binaries, some of the rarer classes and disputed evolutionary scenarios will not be touched upon or will be much simplified.

First Things

Star formation is not very well understood, and this is doubly true for binaries. As a rule, dense clouds of gas and dust hide stellar nurseries until the stars have grown bright enough to blow away these relics of their birth. Thus we see protostars only as sources of infrared radiation (and sometimes radio emission from surrounding hot gas) and do not have the clues to composition, motion, and mass that optical spectroscopy can afford. For binary systems, there is an additional problem: The more massive and faster-evolving member of the pair will greatly outshine its prestellar companion once it begins nuclear reactions. Thus we have identified very few binaries in which even one star is still contracting and has not yet reached the main sequence. One possible example is BM Orionis, an eclipsing variable and the faintest of the four Trapezium stars in the Orion nebula.

Only in 1981 did H. Melvin Dyck of the University of Hawaii and his colleagues spot an infrared companion to T Tauri, the prototype of stars that have not yet reached the Main Sequence. This was the first clear example of a pair in which both stars are in their formative stages. Almost nothing is firmly established about the companion as yet, not even its mass. We may expect to find many more such objects in the data from the InfraRed Astronomical Satellite (IRAS) and, in future years, other space-based infrared observatories. An important unanswered question is how and when protostars contracting to become binaries get rid of such different amounts of angular momentum that they end up with a range of separations all the way from nearly a light-year down to the diameters of the stars themselves.

By the time we see most binary systems, both stars have begun nuclear reactions and are normal members of the main sequence, producing energy by converting hydrogen to helium. Wide pairs, where each star completes its evolution oblivious of the other, will not concern us further. Other doubles are so close that the stars share a common envelope almost from the very beginning. W Ursae Majoris and SV Centauri are the prototypes of these contact binaries (of low and high mass, respectively). Even their shapes are distorted by the gravity of their companions, and the shared surface layers permit material to flow freely between them. It now seems probable that most of these systems eventually merge into single, rapidly rotating giant stars, passing out of our story. FK Comae may be an example of such a coalescing binary.

Between the widely separated and the contact systems are pairs whose stars, initially disconnected, come into contact as one or both evolve and expand. These comprise a third to a half of all binaries, including those responsible for Algol-type eclipses, novae, and some supernovae.

Middle Life

The more massive star of a pair—the primary—is always the fastest to evolve and the first to expand away from the Main Sequence. It sometimes reacts to the presence of its companion quite early. RS Canum Venaticorum is the prototype of a class of variables showing bright emission lines, unusually strong radio and x-ray output, and evidence of starspots. The cause, apparently, is the rapid rotation forced on the primary by tidal effects of the secondary. (High surface activity on stars is believed to be due to fast rotation.) Most other radio-bright stars are also binaries, the emission coming from colliding stellar winds and the like.

Around any two orbiting bodies is an imaginary surface forming their Roche lobes (Figure 1). This hourglass-shaped figure marks the limit of the regions where each object holds gravitational sway; matter outside it can be captured by either. As long as a star stays small enough to fit within its Roche lobe, its gravity is sufficient to hold it together. If the star expands beyond this limit, it embarks on a long and tumultuous career that will change it beyond recognition and may end in a colossal fireworks.

As the primary expands toward the red giant stage, it eventually fills its Roche lobe and starts to overflow. Some of the excess gas falls toward the secondary star, often forming a rapidly rotating accretion disk around it. Once this process begins, it quickly accelerates, because as the star loses mass its Roche lobe shrinks. This process continues until so much material has been dumped that the primary is the less massive of the pair. (We will continue to call it the primary, however.) At this point the primary's Roche lobe starts expanding again. The mass transfer happens much more slowly now and lasts much longer, because it is driven only by the primary's attempt to expand still further and become a red giant—while the lobe is also expanding.

We catch only a few stars in the rapid-transfer stage, since it lasts so briefly. Beta Lyrae is a standard example. A larger class, named for W Serpentis, is just completing the rapid phase. The brightnesses and spectra of rapid-transfer stars show fast, erratic variations caused by the stream of gas from the primary hitting the accretion disk and spiraling down to the secondary's surface.

In addition to mass flow between the stars, much gas (and angular momentum) can be lost entirely from the system at this time. We know this because we see systems at later phases whose components are closer together than they could ever have been when both were Main Sequence stars. Material from the primary probably wraps around both bodies in a common envelope, which they fling off like a spoon stirring a cup of coffee too rapidly.

As mass transfer and loss slow down, the system resembles the famous eclipsing binary Algol. The cool, giant primary in this system regularly eclipses the hot, Main Sequence secondary—which now has the greater mass. Explaining this seeming paradox, where the less massive star of a pair is more highly evolved, was one of the early triumphs of binary star theory.

The primary continues its evolution and, in due course, ignites its helium core or sheds all of its hydrogen-rich envelope or both. Either event shrinks the star away from its Roche lobe. Mass transfer stops, and the system becomes much less conspicuous.

Eventually, the primary reaches its end state as a white dwarf, neutron star, or black hole, which may or may not be detectable. The Hyades variable V471 Tauri is the prototype of systems where we see both a white dwarf primary and a Main Sequence secondary close enough together that a common envelope must have once surrounded them. A neutron star primary in a system at this stage is unlikely to be detectable as a pulsar, since stray gas from the secondary quenches the pulsed radio emission.

Which stellar end product the primary becomes depends on its original mass and how much it has lost. Stars that started out with up to 10 or 12 times the mass of the sun typically make white dwarfs. Those from 10 or 12 up to at least 50 solar masses become neutron stars, and a few very massive ones may make black holes. The supernova explosion caused by neutron star formation can sometimes disrupt the binary, sending the two stars dashing off in opposite directions. More often, an asymmetric explosion makes the whole system recoil at 100 kilometers per second or more.

A system with a dead primary may reveal its wild past in several ways. First, the secondary will be more massive than a single star of the same age could be. The puzzling blue stragglers may be examples. These members of star clusters are too massive (too blue and bright) to have spent the lifetime of the cluster in their present state.

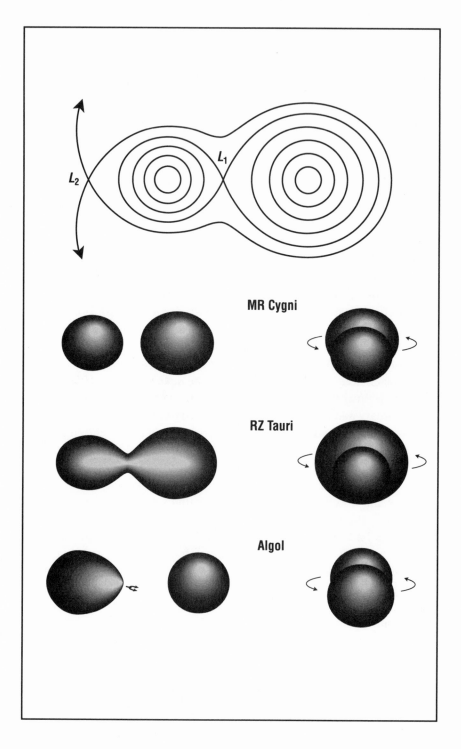

MR Cygni

RZ Tauri

Algol

FIGURE 1. *Roche lobes.*

The gravitational environment in a binary system is much more complex than that near a single star. And some of the resulting effects seem to defy intuition. In considering a gravitational field, it is useful to think of equipotential surfaces, defined by where a particle can move freely without gaining or losing energy. Sea level on earth is a good example of a natural equipotential surface; a ship can sail anywhere on it without climbing uphill or falling downhill. This is true even though the sea departs from a true sphere (due to the earth's rotation and slight irregularities in its mass distribution). When a ship sails from a pole to the equator, it actually moves 14 miles farther from the earth's center—but since it does not have to spend energy climbing, it cannot be said to go uphill.

In the case of a double star, imagine a reference frame rotating at the same rate as the stars so that they appear stationary. Equipotential surfaces in this reference frame are shown in cross section. The two lobe shapes just touching are called the inner critical surfaces, or the Roche lobes. Not far outside these is the outer critical surface, beyond which particles have enough energy to leave the system entirely.

If the stars are relatively small, they will fill the small, separate, spherical or egg-shaped surfaces. An example of such a detached system is the eclipsing binary MR Cygni, whose approximate shape and separation are illustrated here. If both stars are so large that they merge, they fill one of the equipotentials between the inner and outer critical surfaces. Such is the case for the contact binary RZ Tauri. A ship on this rather odd-looking object could sail anywhere on its surface without going uphill or downhill—including from one component to the other across the neck joining them. The stars' surface material presumably circulates just as freely.

If one star is big enough to fill its Roche lobe but the other isn't, then we have a semidetached system and our hypothetical ship is doomed. It can travel anywhere on the larger star's surface, but when it reaches the place where the lobes join, it will sail off the edge of its world—just as in the fantasies of medieval mariners. It will fall through space downward (through smaller equipotentials) toward the smaller star, as in the picture of Algol. This is what stellar material does in a binary undergoing Roche-lobe overflow.

(Computer-generated star pictures are by Robert E. Wilson, University of Florida; courtesy of Mercury.)

Second, either star or both may show evidence of unusual composition, due to the primary having been stripped all the way down to where hydrogen- or helium-burning reactions have taken place. Highly evolved primaries that have not yet become white dwarfs can appear as helium stars (those with more helium than hydrogen atoms in their atmospheres), such as Upsilon Sagittarii and KS Persei. Or they may form Wolf–Rayet binaries, in which a compact, very bright primary with no discernible hydrogen (but lots of helium and nitrogen or carbon) orbits a fairly normal type O or B Main Sequence secondary.

The barium stars may belong in here, too. This element is made by slow addition of neutrons to iron in parts of stars where hydrogen- and helium-burning zones interact. Thus, excess barium shows that a star's surface has been contaminated by products from nuclear burning in a stellar core—its own or a companion's. One class of barium star seems always to be found in binaries with white dwarf companions of fairly low mass. The obvious explanation, that the pollution came from the companion, is not entirely supported by observations, and these systems are still rather a mystery.

Third, portions of the envelope shed by the primary may still be visible, as in the planetary nebulae that have binary central stars. UU Sagittae, FG Sagittae, NGC 246, NGC 1514, and NGC 6543 fall into this category. The supernova remnant G 109.1 − 1.0, whose central object is an x-ray binary, is another example of a recently deceased star surrounded by its ejected envelope.

Second Childhood

Now, at last, the secondary gets its turn. As it evolves and expands, material flows back onto the primary—first slowly as a stellar wind, then in rapid Roche-lobe overflow, then slowly again as the mass ratio is once more reversed. What we observe depends both on the rate of flow and on the kind of object the primary has become.

Gas accreting onto a neutron star or black hole at a slow to moderate rate (10^{-10} to 10^{-8} solar mass per year) gets very hot as it drops down into the primary's deep gravitational well. It emits x rays with a power from thousands to nearly a million times the total luminosity of the sun. The well-known strong galactic x-ray sources, such as Cygnus X-1, Scorpius X-1, Hercules X-1, and Vela X-1, are all of this type.

More rapid transfer of gas produces a thick envelope around the primary, and the x rays do not get out for us to see. Continuing accretion onto a neutron star in this fashion is an obvious way to form a black hole, but it does not seem to happen very often. The primaries of Cygnus X-1 and LMC X-1, two of the best black hole candidates, are both too massive to have grown by this mechanism during the lifetimes of their secondaries.

If the primary is a white dwarf, the flow of gas onto it results in a wide range of phenomena that collectively define the cataclysmic variables. Included are novae, dwarf novae, recurrent novae, novalike variables, symbiotic stars, and polars. All show slight, rapid flickering, and some, of course, explode. Two different energy sources are involved. First, gas dropping down toward the white dwarf releases its gravitational energy as heat and therefore shines in visible and ultraviolet light. Thus, whenever the rate of flow changes, so does the system's brightness. This happens erratically in symbiotic stars and quasi-periodically in the dwarf novae. Second, the accreting hydrogen can release nuclear energy by burning to helium. To do so it must become very hot and dense. If the transfer rate is moderate, hydrogen accumulates on the white dwarf until there is a layer of 10^{-4} solar masses or so. Runaway nuclear fusion then commences at the bottom of the layer. Novae and recurrent novae result from such hydrogen explosions on binary white dwarfs. Any hydrogen not used up is normally blown off in a rapidly expanding shell.

Of the other types, polars are rather like dwarf novae except that the accretion is channeled not into a disk, as is usually the case, but into two columns by the white dwarf's strong magnetic field. The novalike variables behave like old novae but have never been known to explode. They probably include classical nova systems that we simply haven't yet caught in an outburst. There should be quite a few of these, since a nova's explosions come only every 10,000 to 100,000 years. Other novalike variables are probably dwarf novae temporarily stuck at a high flow rate. Much work remains to be done in sorting out the classes of cataclysmic variables, the physics involved, the aberrant types, and the exact progenitors, evolutionary history, and causes of mass transfer for each.

Not all combinations of donor star and flow rate will be noticed. With the lowest possible transfer rate, presumably nothing interesting happens. And at the highest rates, the hydrogen burns steadily as it arrives, so the white dwarf mimics a normal hydrogen-shell-burning giant star. As the newly formed helium accumulates, it too can gradually burn, to carbon and oxygen. If the dwarf collects enough matter to ap-

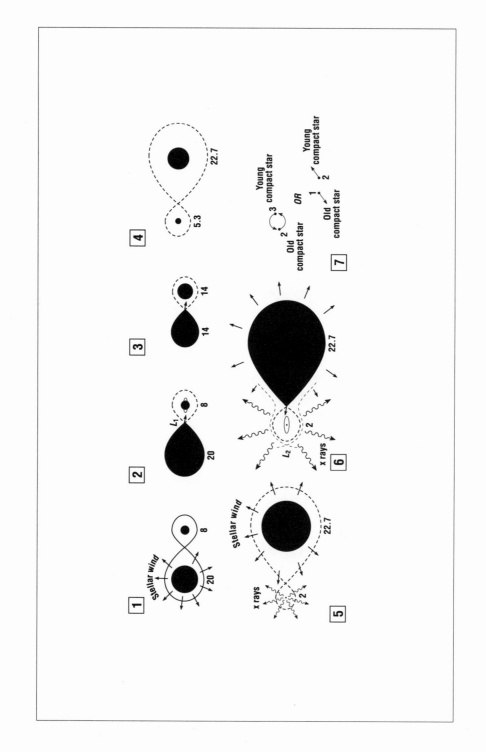

FIGURE 2. *Life and death of a close binary.*

This sequence shows a likely history for the binary x-ray source Centaurus X-3; it illustrates many life stages of interacting binaries in general. The stars' sizes, shapes, and masses (in units of the sun's mass) are taken from Nature Physical Science, *Vol. 239, No. 92 (1972).*

1. As the primary star begins evolving toward the red giant stage, it swells toward its Roche lobe and begins to blow a strong stellar wind throughout the system.

2. When the primary fills its lobe, mass transfer begins in earnest. A curving stream of gas falls from the first Lagrangian (L_1) point toward the secondary, forming a disk of material spiraling down to the secondary's surface.

3. After the primary has dumped enough material to equalize the masses of the two stars, transfer slows down again and a weaker stream flows directly onto the secondary star.

4. Mass transfer from the primary has ceased, leaving only a small helium or Wolf–Rayet star behind.

5. The helium star has exploded as a supernova, blowing most of its material out of the system. The small remnant is a neutron star. The secondary, its evolution speeded up by mass added earlier, is now swelling to the red giant stage and emitting a strong wind. The wind striking the neutron star can turn it into an x-ray source.

6. Once the secondary expands to fill its Roche lobe, rapid mass transfer produces a stream back toward the primary, an accretion disk, and powerful x rays. This is the stage Centaurus X-3 is at today. Mass may be flung entirely out of the system through the L_2 point.

7. The secondary, too, will eventually live out its life to become a white dwarf, neutron star, or black hole. This younger compact star may stay in orbit with the primary or may fly away.

proach a critical value near 1.4 solar masses (the Chandrasekhar limit), it burns its entire helium, carbon, and oxygen supply in one of several violent processes. This is our "best buy" model for type I supernovae, the sort that occur among old stars. The table in "Cataclysmic Variables," immediately following, sums up what we think we know about the many kinds of cataclysmic variables.

Last Things

The secondary, like the primary, has several options for its final state— white dwarf, neutron star, or black hole—and the one reached is determined by its mass once the second transfer phase is over. A white dwarf pair, at first surrounded by a planetary nebula, is the most frequent end point simply because low-mass stars are the most common. UU Sagittae may be such a second-time-around planetary. We see white dwarf pairs with a great range of separations, from the wide visual double G 107–69/70 to AM Canum Venaticorum, whose components are in contact.

Very close white dwarf pairs lose angular momentum via gravitational radiation and can spiral together within the present age of the galaxy. If, when they merge, the total mass exceeds the Chandrasekhar limit, their carbon and oxygen nuclei will ignite and explode, probably making some sort of supernova.

When one or both components are neutron stars and there is not too much stray gas around, we will see a binary pulsar. The orbital motion produces Doppler shifts in the pulse period, revealing the companion even if it is too faint to see directly. Of the three well-known binary pulsars, two probably have white dwarf companions. The third, the famous emitter of gravitational waves PSR 1913 + 16, has another neutron star. Very rapidly spinning but old pulsars may also be products of close binary systems.

Once pulsar emission stops, neutron star pairs become very inconspicuous, as are black hole pairs from the outset. We expect to detect only a final burst of gravitational radiation as they spiral together and perhaps a spurt of heavy elements if a neutron star is incompletely swallowed by a black hole.

Ultimately, then, no close binary lasts forever. If its stars get through their eventful lives without merging or flying apart, their dead, nearly invisible hulks are destined to approach each other very slowly and ultimately join—assuming that the universe itself lasts that long.

Cataclysmic Variables

Acataclysm, according to the *Oxford English Dictionary,* is a
sudden convulsion or alteration of physical conditions (this
vague use of a word from the Greek for deluge being blamed
on the geologists). A cataclysmic binary is an orbiting pair of stars subject to one or more sudden, fairly unpredictable increases in luminosity,
from which we deduce the convulsion and altered physical conditions.
The name was coined by Cecilia H. Payne-Gaposchkin and Sergei
Gaposchkin of Harvard, who also considered calling the systems catas-
trophic binaries. But they made the right choice, because a catastrophe
is something you don't survive; and we now know that these systems
not only survive their assorted convulsions but will undergo them again
and again on various time scales.

Of the various types of cataclysmic binaries (which we will get to
know better in due course) the best known is the classical nova. The
most recent naked-eye example appeared in the constellation Cygnus in
late August 1975. Although it was one of the brightest stars in that constellation for a few nights, by now it is again only to be seen with rather
large telescopes. But don't give up if you didn't see it in 1975—it
should flare up again along about the year 101,975. Apart from a nova
or recurrent nova outburst every decade or two, all cataclysmic binaries
are visible only through telescopes and most are fainter than 10th (apparent) magnitude.[*]

Robert Kraft took a crucial step in our understanding of cataclysmic
binaries in 1964 when he demonstrated that many of them (we now think
all) are binary systems in which one of the two stars is a white dwarf.
White dwarfs are the final stage in the evolution of stars like the sun (that
is, having initial masses of about 0.4 to 6 times that of our sun), in which

---

[*]The greater the magnitude, the dimmer the object. Anything fainter than 6th magnitude cannot be
seen with the naked eye.

all available forms of nuclear energy have been used up. The dying star contracts under the force of its own gravity until the electrons in the gas at its center are crammed together as tightly as quantum mechanics allows. This happens at a density about a million times that of water (or the sun, or you, for that matter). Thus a white dwarf with a mass equal to that of our sun has a radius of only about 7000 kilometers, about like that of the earth.

What difference does all this make? Well, there are two effects. First, you know what would happen if you were to jump from the top of a tall building and hit the surface of the earth. As you fall in the earth's gravitational field, you release the kind of energy called gravitational potential, which you acquired by climbing the stairs to the top. During the fall, this is turned into kinetic energy (you move faster and faster). And when you hit the ground, the kinetic energy goes into heating the ground, making a loud noise, breaking the chemical bonds that hold your legs on, and so forth. But a white dwarf has the same radius as the earth and a mass about 300,000 times bigger, thus a gravitational field 300,000 times stronger. Therefore something falling to the surface of a white dwarf releases 300,000 times as much potential energy as it would falling to earth, and the splat when it lands is awesome to contemplate.

Thus gas (or anything else) falling onto a white dwarf releases lots of gravitational potential energy. The second effect is that, in due course, the gas will also release a great deal of nuclear energy. Seventy percent or more of the gas in the universe is hydrogen. And hydrogen objects to being squeezed to high density. Normally, this doesn't bother a white dwarf—it burned all its hydrogen (and helium) long ago to keep shining over its ten-billion-or-so-year lifetime as a normal star and is now made mostly of carbon and oxygen. But if fresh hydrogen gas flows down onto the star and piles up, eventually the bottom of the layer reaches high density and begins to burn (in the nuclear sense) vigorously and explosively. The gory details of this process are being worked out by teams of astronomers in Japan, Israel, the United States, the Soviet Union, and probably other places. But the result of their calculations indicates that lots of energy should be released in a few days or weeks, and much of the offending hydrogen should be blown back out into space.

An isolated white dwarf, such as our sun will eventually become, could travel through the hard vacuum (about one atom per cubic centimeter) of interstellar space for more than the age of the universe without ever collecting enough gas to make either a splat or a nuclear explosion. But binary systems are different. The companion star (especially if it is not much further from the white dwarf than its own radius) can act as a rich source of gas. And this is just the sort of situation that Kraft found for the cataclysmic binaries: a white dwarf in a rather close orbit around a normal star (usually a main sequence star like the Sun, but sometimes a red giant). Already the discerning reader will see the possibility of several kinds of violent behavior—either splats or nuclear explosions with the gas coming from either Main Sequence or giant donors.

We need to make at least one further distinction concerning how the gas flows from donor to white dwarf. Again there are two possibilities. Nearly all stars (including the sun) have a wind blowing material away from their surfaces all the time. The solar wind is both rather slow moving and rather tenuous. Other stars, especially giants, can have more copious stellar winds, capable of carrying off several percent of the total mass of the star in its lifetime. This is the slower and less efficient way of transferring gas to the white dwarf, because most of the wind never hits it at all but goes off into space. It results in the less spectacular sorts of cataclysmic binaries.

Rapid and efficient mass transfer occurs when the normal star is large enough to fill or overflow an imaginary surface called the Roche lobe (see the figure on p. 140). Gas that touches this surface is free to flow anywhere on it (without gaining or losing any energy), in particular through the point called L_1 in Figure 1, and then down onto the other star. Roche-lobe overflow can transfer more than half the mass of the donor star over its lifetime. Again the transfer is most rapid for giant donors.

The figure shows an idealization of a system with Roche-lobe overflow transfer. There are five critical parts: (1) the normal star on the left, with a funny shape because it is filling its lobe, (2) the white dwarf on the right (the little dot at the center), (3) a stream of gas flowing from the intersection of the two lobes (called L_1, the first Lagrangian point)[*] toward the white dwarf, (4) an accretion disk, consisting of gas in the process of flowing down onto the white dwarf (this takes a while because the gas travels in a sort of spiral path, as implied by the little arrows), and (5) a hot spot, H, where the stream hits the disk.

A theorist would say we are almost through. We can make a little table of all the possible combinations of energy source, type of donor star, and

---

[*]Roche and Lagrange were mathematicians who calculated some of the properties of systems in which two objects orbit each other under the force of gravity.

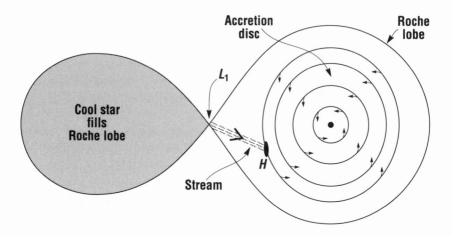

FIGURE 1. *Idealization of a typical cataclysmic binary. A relatively cool (Main Sequence or giant) star is large enough to fill its Roche lobe, a surface on which gas can move freely without gaining or losing energy. Gas at* L₁ *therefore falls in a stream toward the hot white dwarf (the dot at the middle of the right-hand lobe). But the gas has too much angular momentum (rotation) to fall straight onto the star and thus spirals into a disk, from which it gradually accretes onto the star along paths like those indicated by the little arrows. A hot spot, H, is formed where the stream collides with the disk. Some gas transfer also occurs when the cool star is not large enough to fill its Roche lobe, but such systems are less spectacular, and there is normally no disk. Here the white dwarf is the more massive of the two stars (so its Roche lobe is the larger). The opposite case also occurs.*

mass transfer process, give them names, and go home for the day. An observer will say most of the work is yet to be done. Observers have spent decades collecting data on the various types of cataclysmic binaries, and we haven't in the least shown that the systems in our table will look like the systems they observe. Thus we must next take a look at the data and see whether there is any correspondence between the classes of systems in our table and the classes of observed binary systems. (Of course it must all come out pretty well, or I wouldn't be writing this.)

What does the real world look like? Observers typically record for us (a) the brightness and color of a system over time and (b) spectrograms of the light coming from the system at various times. The light curves (the brightness over time) tell us how much of an outburst occurs, how long it lasts, and how long you have to wait between outbursts. Sometimes we also see an eclipse, as the white dwarf (plus gas stream, accretion disk, and hot spot) goes back of the normal star. (That's one good way to tell you really have a binary.) The eclipse also helps us study one star at a time and be sure whether the system contains a white dwarf plus a Main Sequence star or a white dwarf plus a giant. The spectrograms tell us how much gas is around, where it is in the system (e.g., does it get eclipsed), how hot it is, and how fast it is moving (from the Doppler effect of velocity on wavelength of light). Colors and spectra in and out of eclipse tell us whether the companion is a giant or a Main Sequence star and its temperature.

On the basis of light curves and spectra, the observers have separated cataclysmic binaries into the following categories: classical novae, dwarf novae, recurrent novae, nova-like variables (with a subset, called polars, having unusually strong magnetic fields), symbiotic stars, and V471 Tauri stars.* The first eight columns of the table summarize, crudely and not without prejudice, an almost incredible amount of information derived from their work. Now the theorists come back into the picture and ask what sort of mass-transfer mechanism and rate and what energy sources are required to produce outbursts like the ones seen. They thereby fill in the remaining columns of the table.

For instance, Roche-lobe overflow, by definition, can occur only when the cool star fills its Roche lobe. So the theorist must compare the size of the star (known from its temperature and luminosity) with the size of the lobe (known from the size of the binary star orbit). That takes care of the "Cause of mass transfer" column. Deciding about the energy source is a bit less straightforward and requires considering both the amount of energy released in comparison with the gas available (remember each gram of hydrogen has 50 times as much nuclear as gravitational energy to give up) and the length of time over which the outburst dies off.

Luckily, we have some extra checks on whether or not this synthesis of observation and theory has been done correctly. For instance, we know that only nuclear energy is adequate to blow gas back off the white dwarf, and, as we can see in the column labeled "Gas expelled," only the

---

*Astronomers often name a whole class of stars after the first one of its kind to be found.

Observed and Calculated* Properties of Cataclysmic Binaries

System Type	Number known and examples	Stars	Outburst amplitude	Outburst duration	Time between outbursts	Gas expelled from system?	Features seen in spectrum	Cause of mass transfer	Energy sources**	
									Outburst	Quiescent
classical novae	c. 200 DQ Her	MS + WD†	$9-14^m$	months	$(10^4-10^5$ yr)	yes	disk, stream	Roche-lobe overflow	nuclear	gravitational
recurrent novae	5 T Cor Bor	giant + WD	$7-9^m$	months	10 – 100 yr	yes	disk	Roche-lobe overflow	nuclear	gravitational
dwarf novae	c. 300 SS Cyg, U Gem	MS + WD	$2-6^m$	days	10 da – 30 yr	no	disk, stream, hot spot	Roche-lobe overflow	gravitational	gravitational
nova-like variables	dozens UX UMa	MS + WD	irregular variability	—	—	no	disk and/or stream, hot spot	Roche-lobe overflow	—	gravitational
polars	4 AM Her	MS + WD	irregular variability	—	—	no	strong magnetic field disk, stream, etc.	Roche-lobe overflow	—	gravitational
V471 Tauri stars	2 – 3 V471 Tau	MS + WD	—	—	—	no	—	stellar wind	—	—
symbiotic stars	dozens Z And	giant + WD	irregular variability	—	—	no	hot gas	stellar wind	—	gravitational
not seen	—	MS + WD	$(9-14^m)$	—	(very long)	(yes)	—	stellar wind	nuclear	gravitational
not seen	—	MS + WD	(tiny)	—	—	(no)	—	stellar wind	gravitational	gravitational
not seen	—	giant + WD	$(7-9^m)$	(months)	(very long)	(yes)	—	stellar wind	nuclear	gravitational

* Calculated properties are those in the last three columns and those enclosed in parentheses.
** In addition to ordinary light from star surfaces
† Main Sequence star and white dwarf

systems with nuclear-powered outbursts show this characteristic. Only Roche-lobe mass transfer should give rise to a stream and disk, and that is indeed what our table indicates.

We can also use theory to fill in some of the blanks. For instance, after a classical nova outburst, you will have to wait 10,000 to 100,000 years for mass transfer at the observed rate to deposit what the theorists say is enough hydrogen on the white dwarf to explode again. And we are evidently seeing the nova-like variables in the long stretches between nuclear outbursts (though some may be dwarf novae in a semi-permanent state of eruption).

Other connections can be made. As the normal stars in symbiotic and V471 Tauri systems age, they will gradually expand until they fill their Roche lobes, initiating rapid mass transfer. Future astronomers should then see them as recurrent novae and classical novae respectively. The only important difference between the recurrent and classical novae is that giants transfer gas more rapidly. Thus the interval between outbursts is shorter, and some systems have gone off more than once in historic times. We can even understand why a few of the theoretical combinations (the last few lines of the table) don't correspond to observed classes. For instance, wind transfer is so slow that it would take millions of years to build up enough hydrogen for a nuclear explosion, so we just haven't seen one of those yet (but be patient). In addition, also because stellar winds transfer relatively little gas, the amount of gravitational potential energy liberated isn't enough to make an outburst, but only a flicker. Finally, Roche-lobe overflow from a giant puts out such a flood of gas that it flows smoothly onto the white dwarf and contributes to the average total light of the system rather than making outbursts.

Two questions inspired by the table remain: (1) what happens to the nuclear energy of the gas deposited on the white dwarf in dwarf novae? and (2) what happens to the gravitational energy of the gas in classical and recurrent novae? The first is easy: hydrogen accumulates on the white dwarf and every 100,000 years or so gives rise to a nuclear (classical nova) outburst. Since a few hundred dwarf novae are known at present (mostly the nearest ones out of a total galactic population of a million or so), there is a 10-percent chance that one of them will go off in your lifetime (if you give up smoking, exercise regularly, etc.). The second question sounds easy too, but leads us into a morass. The gravitational potential energy is the smaller of the two energy sources by a factor of about 50 (which is why nova outbursts are so much brighter than dwarf nova ones). Thus we believe that the gravitational potential energy is simply released as gas flows from stream to disc to star and contributes to the average (nonburst) light that we see from the system.

OK as far as it goes—but it doesn't go quite far enough. We need to know why mass deposition on the white dwarf occurs at a constant rate in the classical and recurrent novae but unsteadily (giving rise to the outbursts) in dwarf novae. The average rate seems to be smaller in the dwarf novae, and this must have something to do with the problem, but we don't quite know what. Two possibilities pervade the literature on the subject: the instability suggested by Geoffrey Bath, which occurs in the outer layers of the cool, donor star and turns the gas flow on and off at L_1, and the instability suggested by Osaki and Paczyński, which occurs in the outer part of the accretion disc and changes the rate of fall onto the white dwarf. Partisans of one remain on speaking terms with partisans of the other, but only barely.

Some readers may still be bothered by two things. One is whether or not to believe all this. For those with some background, it might be best to find the original literature presenting the observations and discussing the theoretical models. The easiest place I know to start is with a pair of review papers, rather more technical than this one, in volume 19 of the *Quarterly Journal of the Royal Astronomical Society*. They were written by G. T. Bath and by A. D. Mallama and V. L. Trimble and will refer you to some of the research literature and yet more technical reviews.

Finally, we might ask ourselves how a pair of stars ever gets itself into a mess like this and what variations on the theme in Figure 1 are possible (suppose, for example, we imagine replacing the white dwarf with a neutron star). This takes us into the whole broader subject of the evolution of close binary systems, which requires some thousands of additional words for its explication. It is addressed in the article "Close Binaries" in this volume.

DEATHS OF STARS

White Dwarfs:
The Once
and Future Suns

A hundred billion years or so from now, the astronomers who study white dwarfs and other faint stars will inherit the skies by default, for nothing else will be left shining. In the meantime, it may seem a bit difficult to get excited about a class of objects not even one of which is visible to the unaided eye. Nevertheless, recent conferences devoted to white dwarfs have attracted several hundred of the world's 7000 professional astronomers, so there must be something interesting going on.

Perhaps the most obvious drawing card is that white dwarfs mark the endpoint of the evolution of low and intermediate mass stars. Thus, by looking at them, we look five billion years into the future of our own sun. At the same time, because they shine, however dimly, more or less forever, counts of faint white dwarfs enable us to probe the distant past and to ask when stars began to form in the disk of our galaxy.

Finally, and probably of most interest to the practicing astrophysicist, white dwarfs are a laboratory of physical conditions that we cannot duplicate on earth. Their densities exceed those of osmium and platinum by factors of 10^5 and more. Some have magnetic fields of more than 10^8 gauss, where the earth average is about 1 gauss and the most intense artificial fields barely reach 10^6 gauss. And, when binary companions drizzle hydrogen gas down onto them, they respond with violently explosive versions of the nuclear reactions that are mere whimpers in normal stars, perhaps even with the most violent stellar events of all, supernova explosions.

History and Properties

The story of the discovery of white dwarfs and their structure is part of the folklore of modern astronomy. We tell our students of Wilhelm Friedrich Bessel (as in the Bessel functions of mathematics), who analyzed several decades of observations of Sirius and Procyon and, in 1844, concluded that each must have an invisible companion of roughly its own mass. Then comes the great American telescope maker, Alvan Clark, who, testing a newly ground 18-inch objective in 1862, turned it, by chance, toward Sirius and saw a faint dot of light a few seconds of arc from the blaze of the Dog Star. He had discovered Bessel's "invisible" companion, though a possibly apochryphal part of the story has him first suspecting a reflection in his telescope and devoting several days to additional polishing and aligning. Next, in 1914–15, the spectroscopists Walter S. Adams and Henry Norris Russell found that the faint companions of Sirius and 40 Eridani displayed spectra dominated by hydrogen lines. These implied high surface temperatures, such as would normally be found only among much brighter stars, and required the companions to be not much larger than our earth, despite their star-sized masses.

Finally, we invoke the name of S. Chandrasekhar, who, as a young student en route from India to Cambridge in 1930, wrote down equations for the structure of white dwarfs, correctly incorporating both quantum mechanics and relativity. These equations and his 1935 general relativistic models have stood the test of time. They are confirmed by assorted measurements of white dwarf masses and sizes, including the gravitational red shift that occurs when light must climb out of a strong gravitational field. This was finally measured for Sirius B in 1971 by Jesse L. Greenstein, J. Beverly Oke, and Harry L. Shipman at Palomar Observatory.

If you picked a typical white dwarf from the thousand or more that have been catalogued, it would have a surface temperature near 15,000 K (almost three times that of our sun) but less than 1 percent of the solar luminosity, because the area available to radiate is smaller by a factor near 7000. These quantities are directly determined from white dwarf colors and spectra (which tell us the temperature) and parallaxes (which tell us their distances and so their real brightnesses). Other measured properties include an average mass near 60 percent of the sun's, a typical rotation period of an hour or so, and a surface composition of exceedingly pure hydrogen (or, more rarely, very pure helium, and still more rarely H and/or He slightly polluted with carbon, calcium, or other heavier elements). All these we learn from the wavelengths, intensities, and shapes of absorption lines in white dwarf spectra. A few masses in the range of 0.3 to 1.4 M_O come independently from binary star orbits.

Most catalogued white dwarfs are single stars or have only distant binary companions (not suitable for mass determinations). Most have magnetic fields not much greater than those in sunspots (10^4 gauss) though a few percent exceed 10^6 gauss. These strong fields also show up through their effects on spectral line profiles.

Computer models, based on these observed quantities, Chandrasekhar's equations, and large chunks of the rest of physics and astronomy, reveal that our representative star has an interior core of carbon and oxygen, in which no further nuclear reactions will ever occur. Thus a white dwarf continues to shine only because that core is still hot and gradually cooling. The average luminosity implies a few hundred million years of cooling since the parent star of 1 or 2 M_O completed its nuclear evolution, with billions of years of cooling still to come.

Did that last sentence bother you a little? It should, if you happen to recall that the "best guess" age for our galaxy is more than ten billion years. Why is our "average" white dwarf apparently so young? The problem is a particular case of the phenomenon of observational selection effects that bedevils all of astronomy. The youngest white dwarfs are the hottest, hence the brightest. Thus we can see them out to the largest distances, and they end up being heavily over-represented in the catalogues. The truly average white dwarf, found in samples carefully selected to avoid this bias, is a cooler, fainter star than our typical catalogue entry and has been fading slowly for a couple of billion years.

Observational selection is also to blame for the preponderance of single stars in the white dwarf catalogues. Most stars come in pairs, often close enough for the two to exchange material. But even a small hydrogen-burning star will outshine a white dwarf and keep us from seeing it unless the two are well separated. Cataclysmic variables (novae and their cousins) are the exception that proves the rule. They are close binaries, but material flowing onto the white dwarf brightens it to a solar luminosity or more, and we occasionally see them even as far away as other galaxies.

Formation

Which stars make which white dwarfs, and how do they do it? As usual, we must call upon a combination of observations and calculations to sort things out. A stable white dwarf has an enormous gravitational field,

100,000 times as strong as the one we experience on the earth's surface. This must be balanced by internal pressure—not the ordinary thermal pressure of an ideal gas, for these stars have no nuclear energy sources and would contract if only thermal pressure opposed gravity. Instead, the pressure source is a purely quantum-mechanical effect, related to the Pauli exclusion principle and called degeneracy (a description of electron velocities, not moral principles!). The strength of degeneracy pressure is limited by general relativistic effects and by the tendency of electrons and protons to react and form neutrons at high density. The result is a firm upper limit of 1.4 M_O to the mass of a stable white dwarf, generally called the Chandrasekhar limit.

Can we safely conclude that only stars up to 1.4 M_O become white dwarfs? Definitely not, though elementary textbooks often leave that impression. The observations tell us something very different, as follows: The more massive a star is, the faster it exhausts its hydrogen fuel and leaves the Main Sequence. Thus the amount of Main Sequence left in a star cluster tells us both the age and the masses of the stars that have already died. And a handful of young clusters, including NGC 2287 and 2516, the Hyades, and the Pleiades, contain white dwarfs, though the stars just now leaving the Main Sequence range from 3 to 6 M_O! This means that stars at least up to 6 M_O are able to strip themselves down below the Chandrasekhar limit.

Part of the process we see. Many of the very hottest, youngest white dwarfs are ringed by gas clouds called planetary nebulae (because they look a bit like Uranus or Neptune when seen through small telescopes, not for any more profound reason). These nebulae comprise part of the outer layers of the parent star, ejected by pulsation and radiation forces as the stars use up the last of their hydrogen and helium fuels. Expanding at about 30 km/sec, they dissipate in 10,000 years, leaving their cores to fade.

Most planetary nebulae, however, contain only a few tenths of a solar mass of gas. Additional mass loss is therefore needed to get typical 1 or 2 M_O stars down to 0.6 M_O cores, not to mention 6 M_O down to less than 1.4 M_O. Stellar winds throughout the giant phases of stellar evolution provide this additional loss. Extra spectral lines, produced by the moving gas, reveal the existence of these winds, though not, with any great precision, their intensity. This is still under active debate, some astronomers having suggested that winds might strip even 8 to 10 M_O stars to white dwarfs.

Too much mass loss is, however, just as bad as not enough. If stars much above 6 M_O are eroded away, there will be too few massive ones left to process fuels from carbon and oxygen up to iron, collapse as Type II supernovae, and make the pulsars we see. Since little stars are much commoner than big ones, even the intermediate mass range can present a statistical problem if it takes too big a progenitor to make a typical white dwarf.

I cherish an abiding prejudice in favor of relatively weak winds. It derives from some early work of my own that found a larger average white dwarf mass than the one now generally accepted, and has recently received a bit of confirmation. G. Bertelli, A. F. Bressan, and Cesare Chiosi of Padua have recomputed evolutionary models for intermediate-mass stars, using an updated set of nuclear reaction rates and including a form of mixing called semiconvection. They find that peaceful carbon ignition (and thus eventual evolution to a neutron star) can occur in stars as small as 6 M_O and that stellar winds are not quite as strong as has sometimes been advertised. Rotation and magnetic fields remain to be included in future models.

Meanwhile, it seems consistent with calculations, observations, and statistics to say that stars up to 6 M_O form white dwarfs, at a rate of about one per year in the Milky Way. Those above 1 M_O are sufficiently bright and hot to illuminate detectable planetary nebulae during the formation process. And, of the products, most will be carbon–oxygen white dwarfs, the lowest-mass ones helium WDs, and the heaviest will be made of oxygen, neon, and magnesium.

Binary White Dwarfs

Elsewhere in this volume, "Close Binary Stars" addresses how white dwarfs form and evolve in the presence of a companion. We believe with some confidence that such close pairs are responsible for the stellar explosions called novae, and suspect they may also produce one class of supernova, the Type I's, found among older stars.

The more massive star in a pair always evolves first. If it becomes a white dwarf, then, later, when the companion evolves, hydrogen-rich gas inevitably flows onto the surface with its strong gravitational field. The deposition of the gas already releases enough energy to brighten the white dwarf considerably. These are the dwarf novae and symbiotic stars. But as the hydrogen builds up, it gets hot and dense and approaches the conditions required for nuclear fusion. At flow rates near 10^{-9} M_O/year, fusion begins suddenly and explosively after

10^{4-5} years of gradual accumulation. The star brightens by about ten magnitudes in a few days, and the unburned hydrogen is blown off. These are the novae, a few of which are spotted in our galaxy every year. After the fireworks are over, hydrogen begins to accumulate again, and the explosion will repeat in another 10^{4-5} years.

At other flow rates, the hydrogen can burn peacefully, gradually increasing the mass of the white dwarf toward the Chandrasekhar limit. This, too, can trigger an explosion. But now the entire body of the star burns, from helium, carbon, and oxygen up to iron. This much larger energy release suffices to disrupt the star completely, the fragments radiating as much as 10^{10} solar luminosities for a few weeks. At least some of the supernovae that occur in elliptical galaxies and the halos of spirals are thought to be this sort of event.

The systems that blow off hydrogen layers as novae cannot also be building toward the Chandrasekhar limit and supernova explosions. Thus, there may not be enough Type I supernovae produced this way to account for all we see. An alternative scheme with a second white dwarf as donor attracted considerable attention over the past few years as a result. Unfortunately, more detailed calculations suggest that the accreting star collapses rather than detonating. This can release considerable energy, but it will not look like a Type I supernova. The exact progenitors of SN I's remain something of a mystery.

Triumphs and Tribulations

Strong Magnetic Fields

The discovery in the early 1970s of circular polarization implying a field of at least 100 million gauss on the surface of one white dwarf made it immediately clear that we were seeing a new physical regime. The particular star involved (Greenwich +70° 8247) was already famous for having absorption lines in its spectrum that had defied interpretation since Rudolph Minkowski first photographed them at Mt. Wilson in 1939.

The report of the strong field quickly triggered a flurry of papers attempting to identify the lines. The basic idea is that strong magnetic fields grossly perturb electron orbits, so that spectral lines split into many components, shifted by different amounts from their normal wavelengths. Some suggestions were intriguing (transitions in unbound He_2

molecules, for instance), but none matched the observed lines well and none won general acceptance. Sufficiently accurate quantum-mechanical calculations of the expected wavelengths were difficult, and there could be no laboratory data to check them against. Another problem was that the predicted wavelengths changed sharply with field strength, so variations in field strength across the star would smear features beyond detectability.

Two developments finally sorted out the mess. The first was Jesse L. Greenstein's detection in 1983 of an ultraviolet line in the spectrum of Grw +70° 8247, which he attributed to the simplest of all atomic transitions (ground state to first excited state of hydrogen, called Lyman alpha). Thus calculations became more tractable.

The second advance was the development in Tübingen of the idea of stationary components—lines whose wavelength stays nearly constant over a range of field values (owing to a turning point or inflection point in λ vs. B). This permits absorption to be concentrated in a single, narrow wavelength band over the whole stellar surface.

As a result, observers and theorists (including W. Rösner and G. Wunner of Tübingen and Robert F. O'Connell of Louisiana) reached general agreement, attributing both optical and ultraviolet lines in Grw to hydrogen. The lines include components of Lyman alpha and of Balmer alpha and beta, some never seen in the laboratory because they exist only in the presence of a strong magnetic field. The surface field required for Grw is in the range 2 to 6×10^8 gauss. It can never be determined more exactly because of the stationary line phenomenon, but a mystery dating back to 1939 has been solved.

Pulsating White Dwarfs

The commonest white dwarf variables are called ZZ Ceti stars. They display atmospheres of pure hydrogen, surface temperatures near 12,000 K, and pulsation periods of 200 to 1200 seconds. These can be modeled as shape distortions, driven by hydrogen ionization and restored by gravitation (called nonradial g modes). The surprise is that detailed matching of the driving force with pulsation properties requires that the hydrogen atmosphere be exceedingly thin. Until recently, it had to be impossibly thin, 10^{-7} $M_\odot$ or less. No star could burn or eject its hydrogen quite this efficiently. Arthur N. Cox (Los Alamos) and his colleagues have succeeded in modifying this to a less extreme 10^{-4} $M_\odot$. And a new physical process, called diffusive hydrogen burning by its inventor, Georges Michaud (Montreal), can meet the new demands. Let's save the details for the composition discussion.

A second class of pulsating white dwarf, with higher temperatures and nearly pure helium atmospheres, is driven by a helium ionization zone and does not seem to present any special problems. Much excitement has centered on a third class, one of whose half-dozen members is still surrounded by a planetary nebula and so must have completed nuclear reactions very recently indeed. These stars have surface temperatures of 100,000 K or more, arguably the highest known for any kind of star, and their spectra show lines of helium, carbon, and oxygen. The class is named for its prototype, PG 1159-035 from the *P*alomar survey carried out by Richard *G*reen (now at Kitt Peak).

These stars will evolve quickly by both cooling and contracting. The former will lengthen the pulsation periods, the latter shorten them. Thus the discovery by Donald E. Winget (Texas) and his collaborators that the period of PG 1159 is getting shorter at a rate of about 10^{-11} sec/sec means that contraction is currently winning and will change the star appreciably in less than a million years. Data now being collected should tell us also the masses of these stars (probably larger than the general run of WDs) and the extent to which nuclear reactions still contribute to their luminosity.

Most unexpected was the composition needed to match the pulsation periods, just about half oxygen almost up to the surface. The ratio of oxygen to carbon made when helium burns rises with the burning temperature and so should be high only near the center or in very massive stars. But, just as we began to worry, William A. Fowler announced a new, larger cross section for the reaction $C^{12} + He^4 \rightarrow O^{16}$, measured at Münster and Caltech. This increases oxygen production at all temperatures, and the PG 1159 stars can perhaps be regarded as independent confirmation of the laboratory result!

Surface Compositions

Most white dwarfs show strong spectral lines of only one element—hydrogen or helium or carbon. This has been understood to first order since 1945, when Evry Schatzman (Paris) pointed out that their very strong gravitational fields would force the lightest element present to float to the top in much less time than the stars take to evolve. A few white dwarfs show hydrogen plus helium or some heavier elements—carbon, calcium, and magnesium are typical. They too are probably comprehensible. The surface layers reflect a combination of gas freshly accreted (from surrounding interstellar gas or the wind of a companion star) with Schatzman's gravitational settling and radiation pressure (which holds up some metals more effectively than others). Gary Wegner (Dartmouth) has concentrated on a third group in which nearly pure helium atmospheres harbor some carbon (less than we see in the sun, but more than should have survived gravitational settling). He finds that these stars have just the surface temperatures at which currents of rising and falling gas can penetrate deepest into the star, dredging up material from the interior. And we expect the interiors of most white dwarfs to be at least half carbon.

The really difficult case consists of the few stars having helium atmospheres contaminated by elements heavier than carbon (again, less than we see in the sun but more than should have survived gravitational settling). If the explanation is dredge-up, then nuclear reactions in the parent stars must have proceeded much further than we think they can in the sorts of stars that make white dwarfs. And if accreted interstellar gas is to blame, where has the hydrogen gone?

Michaud's diffusive hydrogen burning may provide the answer, as follows. Gravity is trying to float any available hydrogen atop everything else. But there will be an exponentially decreasing tail of hydrogen concentration down into the star. For surface temperatures between about 10,000 and 30,000 K, this tail penetrates into carbon-rich regions hot and dense enough for hydrogen to burn by the carbon–nitrogen–oxygen cycle. So efficient is the burning that it tries to establish an equilibrium hydrogen concentration even smaller than the exponential tail provides. Thus hydrogen is continuously drawn down and burned, leaving the atmospheric abundance below detectability in some stars. Gravitation is tugging downward at the same time on the metals as they accrete, but enough atoms pass through the atmosphere at any given time to make the lines we see. Diffusive hydrogen burning therefore explains an otherwise incomprehensible mixture of elements in a few white dwarf atmospheres. It may also be the reason that ZZ Ceti stars have such thin envelopes and that helium-dominated atmospheres become commoner as white dwarfs age (I forgot to mention that before, but it's true).

Theorists who study white dwarf atmospheres will, however, have their work cut out for them, incorporating no less than five processes in their models. Some combination of gravitational settling, accretion, dredge-up, radiation pressure, and diffusive hydrogen burning may well be able to match every star we see, but a good many details remain to be worked out.

Rarity of Faint White Dwarfs

The cooler a white dwarf gets, the less it radiates and the longer it takes to cool further. Detailed models confirm our resulting expectation that there should be lots more cool, faint WDs than hot, bright ones. And so there are, down to about 10^{-4} $L_\odot$. Then, quite suddenly, there are none (one, vB3, at $M_v = +15.5$, and none fainter). Of course, finding faint stars isn't easy. They make only little spots on our photographic plates, even when very close to us, and are not kind enough to come labeled (except subtly by color and spectral lines) so that we can distinguish them from other little spots. But James Liebert (Arizona) and other experienced observers believe that the rarity is real and needs to be explained.

Explanations so far suggested outnumber the known faint white dwarfs by at least 10 to 1. Some are essentially statistical (wrong choice of the average distance of the stars from the galactic plane, for instance) so that the very few faint ones we have counted represent a sizable population. Others invoke interesting physics to change the rate at which white dwarfs (or at least white dwarf models!) cool. Either slow cooling at high temperatures or faster cooling at low temperatures will work to keep us from seeing stars at $M_v = +15$ to $+17$ (and fainter than that we cannot now look). Some possibilities are (a) changes in the temperature at which the carbon–oxygen core crystallizes (yes, it does; did I forget to mention that, too?), (b) increases or decreases in the rate at which the atmosphere lets interior heat leak out (called opacity), and (c) unexpected extra sources of energy from accreted gas, diffusive hydrogen burning, or chemical fractionation. Each of these has appealed to someone sufficiently to invest some months of his career in modeling it.

But for most of us, the most intriguing possibility is that we may be probing the beginning of star formation in the galactic disk. The faintest white dwarfs, with cooling times approaching the age of the galaxy, must necessarily have come from stars that did not spend much time on the main sequence—that is, from the most massive possible parents. They are much rarer than solar-type stars at any epoch. In addition, massive stars are expected to make white dwarfs of 1 $M_\odot$ or more. These will take about twice as long to cool as 0.6 $M_\odot$ modern white dwarfs because they have a larger energy store and a smaller surface to radiate it from. Thus the oldest WDs could be hiding around $M_v = 15$ along with much younger ones.

It is also just possible that we may be trying to look back to a time before any intermediate-mass stars had formed in the disk. The disk age implied unfortunately depends on getting all the physics details right and could be 6 or 16 billion years or anything in between. We might distinguish the astronomical from the physical explanation by counting the faintest white dwarfs in globular clusters whose ages we know exceed 10 billion years. This will not be easy even with the Hubble Space Telescope as originally designed. Three of the observers who are guaranteed time (because they helped develop it)—John Bahcall (IAS, Princeton), James Westphal (Caltech), and Ivan King (Berkeley)—had intended to try. Sadly, this is one of the many programs rendered impossible by the error in shape of the HST primary mirror. Thus sorting out this particular white dwarf puzzle by this method will have to wait until at least 1994.

The Crab Nebula and Pulsar

"**O**n a chi-chhou day in the fifth month of the first year of the Chih-Ho Reign-Period a 'guest-star' appeared at the southeast of Thien-Kuan, measuring several inches. After more than a year, it faded away." So said the Chinese historian Toktaga in his *Sung shih,* or history of the Sung dynasty.

The people who have studied such things tell us that the day chi-chhou of the fifth month of the first year of the period Chih-Ho was July 4, 1054 A.D., and that Thien-Kuan is near the star Zeta Tauri. It is not quite clear how the ancient Chinese measured inches in the sky (although an inch, held at arm's length, is about $2°$), but if we look today—or preferably tonight—a little more than a degree away from Zeta Tauri, we see a rather unspectacular faint (about 9th magnitude) nebulosity called the Crab Nebula.

It may seem a bit surprising that, out of many thousands of nebulae in the sky and almost 600 temporary stars reported by the Chinese, two can be associated with any degree of confidence. But several factors help. First, the series of Chinese, Japanese, and Korean observations tells us that the object could be seen in the same place for more than 600 days; hence, it was neither a comet nor a meteor. Nor could it have been a variable star or ordinary nova, because there is no suitable remnant star in that part of the sky now. Thus, the 1054 event must have been a supernova, a rare (only about one per century occurs in our entire galaxy) and violent kind of stellar explosion, which can blow off more than half the material of the star involved or even disrupt it completely. One of the Chinese reports mentions that the guest star was "visible by day, like Venus," for 23 days. Since we know roughly how bright a supernova really is, this allows us to estimate the distance to the 1054 event as 4000 to 8000 light-years.

Modern observations of the Crab Nebula, on the other hand, clearly show that it is one of a class of gaseous nebulae called supernova remnants. Its spectrum of emission lines superimposed on a continuum of synchrotron radiation (the kind of radiation emitted by high-speed electrons moving in a magnetic field), combined with the absence of any conspicuous central star, does not admit of any other interpretation. Furthermore, comparison of two pictures of the Crab Nebula line emission features, taken at different times, shows that the nebula is expanding. We can use that measured expansion in two ways: first, when combined with radial velocities measured from the spectrum, it tells us that the object is about 6000 light-years away. And second, if the expansion is extrapolated back in time, it turns out that the nebula would have been a point in about the year 1140 A.D. When we allow for the expansion having been accelerated by the pressure of high-speed electrons and magnetic fields (precisely the ones required to explain the synchrotron continuous spectrum), the most probable date for the explosion is perhaps 100 years earlier. Thus the Crab Nebula can be shown to be the remnant of a supernova which occurred in the same part of the sky, at about the same distance from us, and at about the same time as the event seen by the Chinese.

The connection between the 1054 event and the Crab Nebula was first suggested in the early 1920s and confirmed by about 1940. Since that time, three or four other supernova remnants have been associated with observations of events that produced them in historic times. But no other object has been so well studied or provided so many surprises as the Crab Nebula, and it remains unique in the wide variety of phenomena exhibited. It has features that move, perhaps periodically, at an appreciable fraction of the speed of light; it is an x-ray source as well as a radio source; the polarization of its optical and radio emission shows that they are produced primarily by the synchrotron mechanism—hence, the nebula must contain both a magnetic field and relativistic electrons. There may also be relativistic protons and heavier nuclei present that, if we detected them near the earth, we would call cosmic rays. It was, in fact, the first radio source outside the solar system to be identified with a particular object, the first non-solar-system x-ray source to be so identified, and the first object for which synchrotron radiation (now thought to be the emission mechanism for a wide variety of objects, inside our galaxy and outside it, including quasars) was found to be important.

The nebula also contains a pulsar—the first pulsar to have been found at optical and x-ray, as well as radio, wavelengths. Changes impressed on the radiation from the Crab Nebula pulsar (NP 0532) by the material through which this radiation passes on its way to us even tell us something about the density of the interstellar medium in our galaxy. NP 0532 was the first pulsar to be observed, although the observers did not know what they were seeing. The radio emission from the pulsar was first recorded in Cambridge in 1964 as a small low-frequency radio source located near the middle of the larger radio source corresponding to the nebula. The optical emission from the pulsar looks just like an ordinary 16th-magnitude star, which must have been seen by many people even before the turn of the century. It is only when one makes use of electronic light-sensing devices with good time resolution that it becomes obvious that this star is, in fact, emitting all its light in short bursts separated by about 0.033 second. (This period is gradually increasing, with a time scale for doubling of about a thousand years—yet another link with the 1054 event.) If the period of NP 0532 were even a factor of 2 longer, the light variation could be seen directly. Perhaps it is just as well that it cannot be; 50 years ago an astronomer who came home from an observing run with the report that he had seen a faint star which flickered regularly might well have been suspected of drinking something stronger than water with his midnight lunch.

It is a good omen for astronomy that this melange of observations of different kinds can be put together into a reasonably consistent picture which is in good accord with our theoretical understanding of supernova explosions and their remnants. Even more propitious is the fact that much of the scenario was proposed as far back as 1939, when Fritz Zwicky suggested that the vast energy of a supernova explosion was derived from the formation of a neutron star and, therefore, that there ought to be a neutron star in the center of the Crab Nebula. The very short pulsation period of NP 0532 implies that it is almost certainly a rotating neutron star, a striking confirmation of Zwicky's prediction.

The story starts with an ordinary star of five or ten times the mass of our sun, burning nuclear fuels. Initially, the star burns hydrogen to form helium. Eventually, all the hydrogen in the center of the star is used up, leaving an inert helium core, surrounded by a thin shell in which hydrogen is still being burned. The inert core must contract and heat up, living on its gravitational energy, in order to balance the radiation from its surface. In due course, the center of the star becomes hot enough to enable the helium to burn and form carbon and oxygen, until it, too, is exhausted at the center. Meanwhile, the hydrogen-burning shell continues to eat its way outward through the star. The inert core of carbon and oxygen, in turn, contracts and heats up until further burning occurs.

Thus a highly evolved star will consist of a number of layers, some inert and some burning a variety of nuclear fuels; the central core is very hot and dense, indeed, for a cubic inch of its material would weigh several tons and would contain enough heat energy to boil almost a billion gallons of water.

But this process cannot continue forever. Eventually, either one of the fuels ignites explosively (this probably blows the star apart completely), or the central core is processed by successive nuclear reactions all the way to iron, and no further reactions can extract energy from it. The core is then free to collapse rapidly, liberating large amounts of gravitational energy. While the star was evolving and growing hotter and denser at the center, its outer layers were expanding and cooling. Thus the ultimate explosion effectively occurs inside a large, cool gas cloud. Radiation trapped inside such a cloud is unstable, in the sense that instead of pushing the whole cloud out uniformly, it will break through in channels, forcing the gas into filaments between the channels. So, at this stage, we have (1) radiation that has leaked out and will, in due course, be observed as the supernova event, (2) a few solar masses of gas broken up onto filamentary structures and flying off with velocities of 1000 to 10,000 kilometers per second, and (3) a one or two solar mass core which has collapsed to such a high density that the electrons and protons in it are squeezed together to form neutrons and which is, therefore, called a neutron star. Such a star, once formed, is stable, and cannot contract further even when it cools off.

Most ordinary Main Sequence (hydrogen-burning) stars rotate, with periods of a few hours to a few months. Many of them also have magnetic fields, with strengths ranging from 1 to 1000 or even 10,000 gauss. Angular momentum and magnetic flux are probably approximately conserved when a star collapses. Thus the neutron star may be spinning with a rotation period much less than a second and may have a surface magnetic field as large as 10^{12} or 10^{13} gauss when it is formed.

At this point, the scenario becomes both more complicated and less certain. It is clear that a number of different processes must occur, but there is by no means universal agreement among people working in the field on which processes are most important at which stages. Among the competing mechanisms are gravitational radiation, acceleration of cosmic rays by an outgoing shock wave, acceleration of high-energy charged particles in a magnetosphere around the neutron star, and electromagnetic radiation at the star's rotation frequency. All of these drain rotational energy from the star, increasing its rotation period.

The formation of a neutron star and its early rapid-rotation phase are almost certainly accompanied by the radiation of gravitational waves. Although this radiation may occur as a short burst, it is probably not of the right wavelength to account for the gravitational wave events reported by J. Weber of the University of Maryland. The intensity of a neutron star's gravitational radiation will decline rapidly as the star rotates more slowly and becomes almost spherical. Even for NP 0532, the youngest pulsar known, gravitational radiation is probably negligible.

Cosmic rays, as observed at the earth, are very-high-energy charged particles (mostly protons but also some electrons and nuclei of elements heavier than hydrogen). It is believed that some of these particles have been accelerated to high kinetic energy in supernova events in our own galaxy. But this process must be confined to the early phases of the explosion. Although there are undoubtedly high-energy charged particles in the Crab Nebula at the present time, these will not be released into the interstellar medium until after they have lost most of their energy.

A decade or two after the supernova event, a magnetized neutron star remains, rotating 50 or 100 times a second and surrounded by an expanding cloud of filaments (already something like ten times the size of the solar system). These filaments are mixed with high-energy charged particles and threaded by a magnetic field, partially derived from the interstellar medium swept up as the cloud expands. As the cloud expands, its density drops, and it becomes transparent to most kinds of electromagnetic radiation, from radio on up to x-ray frequencies.

The magnetic field of the neutron star is probably roughly dipolar, like the field of the earth, an ordinary magnetic star, or a laboratory bar magnet. If the axis of the magnetic field is parallel to the star's rotation axis, then electric fields are produced, whose effect is to lift charged particles off the surface of the star into a surrounding magnetosphere and to accelerate them to high energies. If, on the other hand, the two axes are perpendicular, then the star will radiate very intense electromagnetic radiation at its rotation frequency. This frequency is so low that the radiation cannot propagate through ionized gas; so the ions and electrons are swept along by the electromagnetic radiation and again accelerated to high energies, while a cavity around the neutron star is swept clear of gas. The rate at which energy is lost from the star is the same for the two processes, and, of course, if the magnetic and rotation axes are inclined at some angle other than $0°$ or $90°$, some combination of the two processes will occur. Thus energy is lost from the rotating neutron star, in the form of high-energy particles and low-frequency electromagnetic radiation.

As this energy is lost, the rotation gradually slows down. Sudden, discontinuous changes in the rotation period may also occur. Such changes in period (called glitches) have been observed for both NP 0532 and for the pulsar associated with the supernova remnant in the constellation Vela. The glitches may be due to a settling of the neutron star into a more spherical shape (a sort of starquake), to the sudden escape of a large volume of high-energy particles from the pulsar's magnetosphere, or, just possibly, to a planet sweeping close in on an eccentric orbit around the neutron star.

As a result of the highly efficient transformation of rotational energy, a continuous supply of relativistic electrons diffuses through the surrounding gas cloud or nebula. These electrons find themselves in the magnetic field of the nebula. In addition, any low-frequency electromagnetic radiation that is not used up in accelerating particles close to the pulsar will also spread through the nebula, and to a given particle at a given time, this long-wavelength radiation field will look rather like a static magnetic field. Now an electron shot into a magnetic field goes into an orbit which spirals around the magnetic field lines, and it emits electromagnetic (synchrotron) radiation as it goes. This radiation is polarized and has a frequency which depends both on the strength of the magnetic field and on the energy of the electron. Thus electrons with a wide distribution of energies, such as acceleration near the pulsar is likely to produce, will radiate at x-ray, optical, and radio wavelengths.

The realization that a pulsar can provide a continuing supply of high-energy electrons has solved one of the long-standing problems connected with the Crab Nebula. The highest-energy electrons, which radiate x-rays, lose all of their energy in only a year or two. It was, therefore, puzzling that the Crab Nebula, 1000 years old, is still an x-ray source.

With so many different kinds of things—particles, fields, and radiation—present in a supernova remnant, a wide variety of complicated interactions are bound to take place, some of which we understand less incompletely than others. For instance, the emission line spectrum of the Crab Nebula filaments is well explained by the assumption that the gas is kept ionized by the ultraviolet synchrotron radiation produced in the nebula itself. We have already mentioned that the expansion of the nebula appears to be accelerated by the pressure of the magnetic field and relativistic electrons that produce the synchrotron radiation. Among the more puzzling features are the wisps, regions of extra bright optical synchrotron emission near the center of the nebula, which move around with speeds up to 10% of the speed of light, perhaps oscillating with a period of a year or so.

One of these wisps seems to have become suddenly brighter shortly after the large discontinuity in the period of NP 0532 that occurred in September 1969. At about the same time, the electron density near the center of the nebula increased, as did the radio emission of the nebula at a wavelength of about two centimeters. These changes all seem to indicate that a large burst of high-energy electrons was produced at the time of the glitch. These electrons may have burst out of the pulsar's magnetosphere or been blown off a planet, rather like the tail of a comet.

Finally, as a supernova remnant ages, the nebula will merge with the surrounding interstellar medium and disappear. The pulsar's rotation period will have increased to one or two seconds, and it will emit only radio pulses and perhaps weak x-ray radiation, due to interstellar matter falling onto its surface. More than 50 of these long-period isolated pulsars have been observed near us (compared to 2 still associated with nebulae). Their average age is 10^6 to 10^7 years. Thus, if our galaxy has been producing neutron stars at the present rate over its entire lifetime (about 10^{10} years), there should be about 100 million invisible, dead pulsars in the galaxy. These "cinders" would represent less than 0.1 percent of the mass of the galaxy, but they may be responsible for a large fraction of its soft x-ray emission.

You will notice one serious omission in this discussion: no account has been given of the pulsar mechanism itself. It is clear that the precise periodicity of the radio (or, for NP 0532, radio, optical, and x-ray) pulses from these objects is accounted for by their rotation. Because of the very high intensity of the pulses and the small volume from which they come, it is also clear that they must be produced by large bunches of charged particles radiating in some coherent way. And, finally, these bunches must somehow be confined to a small sector on the surface of the neutron star, so that we see their radiation in pulses as that sector sweeps past us like a searchlight beam. But, so far, neither observation nor theory takes us further than this. The gap is not so serious for our understanding of neutron stars and supernova remnants as it might at first seem. Although the existence of the neutron stars proposed so long ago by Zwicky was first demonstrated by their pulsed radio emission, much less than 1 percent of the energy lost by the rotating star goes into that pulsed emission.

Since the explanation of supernova remnants in terms of rotating magnetic neutron stars and their products has been so successful (at least by astrophysical standards), a good deal of effort has recently gone into trying to explain the behavior of quasi-stellar objects and the active centers of galaxies in some analogous way; the Crab Nebula is thereby taken as a prototype for high-energy objects throughout the universe. The case for this analogy is strengthened by the many properties that some quasi-stellars and active galactic nuclei have in common with the Crab Nebula. Among these are:

1. They are synchrotron radiation sources and must, therefore, contain magnetic fields and relativistic electrons.

2. They radiate a large fraction of their energy at short wavelengths and must, therefore, have a large fraction of their relativistic electrons accelerated to very high energies. The Crab Nebula emits about 90 percent of its radiative energy at x-ray frequencies.

3. They have explosive events, which (like the wisps in the Crab Nebula) propagate at speeds close to the speed of light.

4. They have preferential directions, defined by jets sticking out of their centers or by the axes of radio sources, which consist of two components on opposite sides of their parent galaxies. In the Crab Nebula, the wisps always move along the major axis of the elliptical shape of the nebula, and this is also the axis of polarization of the nebula.

5. Their spectra show emission lines, which must come from gas in filaments at rather similar temperatures and densities to those which prevail in the Crab Nebular filaments.

6. At least one quasar (3C 273) probably contains a small low-frequency radio source, and the radio emission from NP 0532 was first detected as such a source.

7. At least one quasar (3C 345) may exhibit periodic variations in brightness, which would be analogous either to the pulses from NP 0532 or to the apparently periodic behavior of the wisps in the Crab Nebula.

8. They exhibit energetic behavior very far away from what appears to be the source of the energy: the diameter of the Crab Nebula is more than 10^{12} times the diameter of its central neutron star. And some radio galaxies have small radio-emitting components separated from their parent galaxies by distances up to 10^4 times the size of the galaxy.

9. Many of them vary in intensity, as does the Crab Nebula at some wavelengths. Two kinds of models are possible: the quasar or active nucleus might be assumed to contain a single, massive (e.g., 10^8 solar masses), rotating, magnetized object or a large number of separate, pulsar-sized objects. The observations do not, at the moment, definitely favor one or the other of these possibilities. But the general idea of rotating magnetic objects acting as sources of low-frequency electromagnetic waves (which can, under some circumstances, travel very great distances without losing their energy) and high-energy particles may be the correct explanation of many properties of active extragalactic objects in the universe.

The Greatest Supernova Since Kepler

Even scientists who have spent the last few years under large rocks cannot help having heard of supernova 1987A in the large Magellanic cloud (the associated neutrino burst having readily penetrated the very largest rocks). It was the first naked-eye supernova since that studied by Kepler in 1604, though if it fulfilled the folklore prediction that when there was another astronomer as great as Kepler there would be a supernova for him to study, we have not yet identified the eponymous astronomer. Radiative fluxes from 1987A are still changing on time scales short compared to publication time scales. Much of what follows must, therefore, be regarded as subject to change without notice.

Incidentally, in accordance with a 1985 resolution of the International Astronomical Union (the body charged with deciding matters of astronomical nomenclature), the supernova is unambiguously 1987A. 1987a was Comet Levy. Supernovae are designated by year of discovery (not name of discoverer) and a capital letter indicating order of discovery during the year.

The Progenitor

About 10 million years ago, one of many small flurries of star formation in the Large Magellanic Cloud gave rise to a few dozen OB stars in and around a region now called NGC 2044.[1] A number of these stars have had their magnitudes and spectral types recorded.[2] One of them, catalogued as Sanduleak −69° 202, appears to have given rise to the supernova. It had two close companions (also blue), and a number of authors suggested early on that a possible fourth red star might have been the actual progenitor.[3]

Several arguments, however, favor Sk $-69°$ 202 itself as the culprit. First, careful photometry and astrometry[4] do not leave much excess light at any wavelength to be credited to a fourth star. Second, IUE spectroscopy during and after the event[5] indicated that, when the supernova itself faded, only two stars remained, separated by the same distance as the previous companions. Third, and requiring lengthier discussion, it has become clear that blue supergiants can produce core-collapse supernovae and that the resulting light curves, spectra, and so forth should look very much like the observations of 1987A.

Combining observations made before the Sanduleak star exploded with its behavior afterwards, we can describe the progenitor[6] as a B3 Ia supergiant with $V = 12.4$, $B–V = +0.04$, visual absorption $A_V = 0.5$, bolometric correction $+1.15$, and distance modulus 18.2–18.8 (corresponding to a distance of 43,000 to 58,000 pc). The bolometric magnitude was, therefore, $-7.8^{+0.3}_{-0.4}$, meaning a luminosity of $4^{+2}_{-1} \times 10^{38}$ ergs/sec. The effective temperature was 15,000 to 16,000 K, and the radius $3 \pm 1 \times 10^{12}$ cm. To achieve this luminosity, the core mass inside the hydrogen-burning shell must have been 6 ± 1 M_O corresponding to a Main Sequence mass of 15 to 22 M_O, and a Main Sequence lifetime of about 10^7 years. Prior to explosion, the star had first mixed significant quantities of helium, CNO-cycle material, and (probably) s-process products into its envelope, and then shed between 3 and 10 M_O of that envelope in a low-velocity, high-density red giant wind, the other 3 to 10 M_O of hydrogen-rich envelope being retained. This relatively dense wind material is now about 10^{18} cm from the star, having been replaced near the star by a higher-velocity, but very tenuous, wind shed during the last 10,000 years or so, when the star had become compact and blue again.

The explosion of a blue supergiant initially caused some puzzlement—perhaps more in the semipopular press than in the astronomical community. In fact, however, even before 1987A, an assortment of evolutionary tracks for massive stars had terminated with the star still blue.[7] Additional tracks of this type have been published since.[8] These stars remain blue through carbon burning (after which interior evolution proceeds so rapidly that the outer layers cannot respond to it anyway), either because of extensive mass loss (Wolf–Rayet stars being the extreme version of this scenario) or because the low metal abundance appropriate to Magellanic Cloud stars prevents radiation pressure from lifting the outer layers. Shklovskii[9] in fact predicted faint SN IIs in irregular galaxies on this basis.

Figure 1 shows a still more relevant evolutionary track for a 20-$M_\odot$ star that spends some time as a red supergiant and returns to blue before exploding. Such loops in the HR diagram are possible because a given set of core and envelope mass and composition parameters does not determine a unique solution to the equations of stellar structure.[10] Rather, there are two stable (and often one unstable) structures possible, and which is followed by a particular model calculation depends very sensitively on composition, mass loss, choice of criteria for convective instability, prescription for handling composition discontinuities left by hydrogen burning, and probably other things.[11] Which path is followed by a particular real star is then a similarly delicate function of the star's real properties—composition (including the amount of helium mixed into the envelope), angular momentum distribution, magnetic field structure, and so forth. Rotationally induced mixing (meridional circulation) may also be a significant factor.[12] Thus we should not be surprised by the simultaneous presence of red and blue supergiants and Wolf–Rayet stars in the vicinity of 1987A, or by the fact that it was one of the blue stars that, on

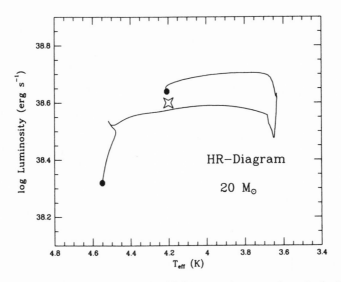

FIGURE 1. *Evolutionary track of a 20-$M_\odot$ star that completes its hydrostatic evolution with a temperature and luminosity similar to that of the progenitor of 1987A (shown as four-pointed star). The model had a metal abundance one-quarter solar, no mixing due to convective overshoot or semiconvection, and convection according to the Ledoux criterion. (From S. E. Woosley, Astrophys. J. 330, 218 (1988); courtesy of Stanford E. Woosley.)*

23 February 1987 (or, rather, some 150,000 to 170,000 years before) reached the point of core collapse and envelope ejection. The news of this event first reached earth in the form of a neutrino burst and, perhaps, gravitational radiation.[13]

The Neutrino Burst and Its Interpretation

Four particle detectors, three of them primarily designed to search for proton decay, each recorded five or more neutrinolike events on 23 February 1987. First detection (and first report), with five counts, came from under Mt. Blanc,[14] about 7.5 hours before optical rise. The detectors in Japan (Kamiokande II, 11 counts),[15] the United States (Irvine–Michigan–Brookhaven—IMB—8 counts),[16] and the USSR (Baksan, 5 counts)[17] recorded their events about 3.5 hours before optical rise, though only IMB had timing accurate to better than 1 minute, and it is an assumption that all three coincide.

Majority opinion, at least in the United States, has decided to believe the Japanese and American reports, to distrust the European event, and to ignore the Soviet one. Theoretical papers and preprints greatly outnumber the 19 counts being interpreted. At least a plurality conclude (a) that the observations are well fit by $6 \pm 3 \times 10^{52}$ ergs in $\bar{\nu}_e$, hence $3 \pm 1 \times 10^{53}$ ergs of total neutrino energy, the binding energy of a 1.4 ± 0.1-$M_\odot$ neutron star, coming out from a neutrinosphere radius of 27 ± 15 km at an initial temperature of 4.2 ± 1 MeV, which cooled in 2 to 5 seconds, and (b) that this is just exactly what we ought to have seen from a core-collapse supernova.[18]

Many of these authors (and others[19]) also find that the near-simultaneous arrival (within 12 seconds) of all the neutrinos, independent of energy, constrains the $\bar{\nu}_e$ rest mass to be less than 10 to 20 eV. This overlaps most laboratory limits and surprises no one. The authors who saw evidence in the arrival times for finite neutrino rest mass[20] seem to have been outvoted.

Several other tight, but inoffensive, limits on neutrino properties follow from the observations. These include a charge less than 10^{-17} that of the electron[21] and a magnetic moment less than 10^{-12} of the Bohr magneton (small enough to rule out solving the solar neutrino problem by helicity flipping in the solar magnetic field[22]). The number of neutrino flavors cannot be more than three.[23] We can exclude an assortment of masses and lifetimes of heavy neutrinos and other particles that couple to or mix with

$\bar{\nu}_e$[24] and a variety of still more exotic processes and couplings.[25] Finally, neutrino oscillation of the MSW (Mikheyev–Smirnov–Wolfenstein) type, widely invoked to reduce solar neutrino detectability, is not ruled out.[26] And the equivalence principle still holds.[27]

One might have hoped that the neutrino burst properties would resolve the longstanding problem of whether ejection from type II supernovae occurs via a prompt or a delayed shock,[28] but the 19 official counts simply do not tell us enough about temperature versus time to distinguish the two.[29]

As an alternative to ignoring the Mt. Blanc counts, it has been suggested that both bursts were real, the second resulting from further collapse to a black hole or from transition to strange quark matter.[30] This choice implies a longer time interval (7.5 versus 3.5 hours) between core collapse and optical rise, which can be an advantage[31] or a disadvantage,[32] depending on who is doing the modeling. A separate serious objection is the very large total neutrino energy implied by the Mt. Blanc counts.[33] A similar energetic objection to the Kamiokande counts may have been resolved by recalculation of the incident neutrino angles.[34] The new distribution is sufficiently isotropic for all 11 events to be induced β decays (rather than one or two direct electron scatterings being required).

The eventual appearance (or nonappearance) of a pulsar will resolve this issue, though not necessarily within the lifetimes of any of the protagonists (especially if neutron stars are not always born with strong magnetic fields).

The Electromagnetic Event

Onset

A capsule statement of our understanding of type II supernovae in general is that the light curves, spectra, and other electromagnetic properties are largely accounted for by a model in which about 10^{51} ergs is deposited (somehow) at the base of an extended red supergiant envelope.[35] Much the same applies to the specific case of 1987A, provided only that the envelope is that of a blue supergiant, compact and with a steep density gradient.[36] A minority view, that $(2-3) \times 10^{51}$ ergs is required,[37] is worth keeping in mind, because, if true, it would almost rule out a delayed shock as the source of SN II ejection.

The main points on which the behavior of 1987A differed from that of normal SN IIs are (a) it brightened in hours rather than days,[38] (b) its peak bolometric luminosity of $9^{+4}_{-3} \times 10^{41}$ ergs/sec[39] was about 10 percent that of a typical SN II (the error bars reflecting both uncertainty in distance to the progenitor and some disagreement between absolute fluxes measured at different observatories; see Figures 2 and 3), (c) its continuous and line spectra in optical, IR, and UV all changed in days rather than weeks, implying very rapid decreases in effective temperature and photospheric velocity with time,[40] and (d) its radio emission also turned on and faded quickly, reaching a peak only 10^{-3} that of previously detected SN IIs, and expanding in less than 48 hours to a size resolved away by VLBI.[41] All these turn out to be comprehensible consequences of the hot, compact envelope structure of the progenitor.

Taking the last point first, radio emission from a handful of previously detected SN IIs has been modeled as the product of ejecta shocking a surrounding circumstellar shell made of the relatively dense, slow-moving wind of a red supergiant. The faint fast character of the

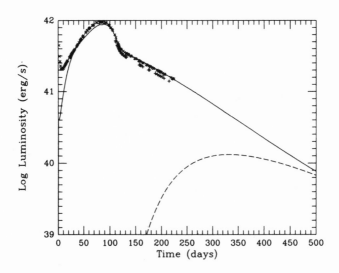

FIGURE 2. *Model light curve for 1987A fitted to bolometric luminosity data from CTIO (lower points) and SAAO (upper points). The model assumes an initial radius of 3×10^{12} cm, an explosion energy density of 0.75×10^{17} ergs/g, the presence of 0.07 $M_\odot$ of Ni^{56}, and some mixing of heavy elements through the hydrogen envelope. The dotted line indicates a prediction of the rate of gamma-ray escape. This particular model includes a 10^{40} ergs/sec pulsar, and it is clear that anything brighter would lead to a contradiction between data and model. (From W. D. Arnett, in* Supernova 1987A in the Large Magellanic Cloud, *ed. M. Kafatos and A. G. Michalitsianos [Cambridge: Cambridge University Press, 1988], p. 301; courtesy of W. David Arnett.)*

1987A emission means that the material available for shocking was much more tenuous than usual, as expected for a blue supergiant wind.[42]

Equally straightforward explanations apply to the first three points, pertaining to light curves[43] and spectra.[44] The shock sent on its way (somehow) by core bounce gets out quickly (having not very far to go and a medium with high sound speed to do it in) but loses a great deal of its energy lifting the envelope out of its deep potential well. Thus, there is less energy left to be radiated. Initial luminosity must briefly have touched 10^{44} ergs/sec, most of which came out at $T \gtrsim 10^5$ K and so in the ultraviolet, ionizing surrounding material[45] and fading in hours.[46] The thinnest outer surface layers received expansion velocities in excess of 20,000 km/sec, but owing to the very steep density profile ($\rho \alpha\ R^{-10}$ or so),[47] we soon saw down to much slower-moving material, already cooled by adiabatic expansion. Hydrogen line velocities, for instance, fell to 2000 km/sec by days 25 to 40 and then fell no

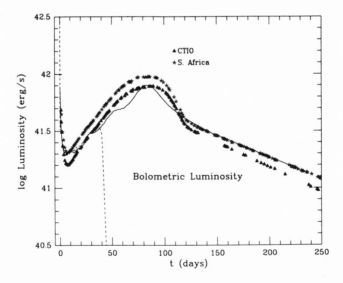

FIGURE 3. *Model light curve for 1987A fitted to bolometric luminosity data from CTIO (lower points) and SAAO (upper points). The model assumes an envelope mass of 10 $M_\odot$ (with a gradient of helium composition through the hydrogen-rich ejecta), an explosion energy of 1.4×10^{51} ergs, and the presence of 0.07 $M_\odot$ of Ni^{56}. The dashed line is the predicted light curve in the absence of the Ni^{56} contribution. (From S. E. Woosley, Astrophys. J. **330,** 218 (1988); courtesy of Stanford E. Woosley.)*

further (indicating that gas still deeper in the star consists of processed material).

Figures 2 and 3 are model light curves based on these ideas. We will return shortly to what happens after the first month. Figure 4 shows the observed optical and UV spectrum at the bottom and three synthetic spectra, representing different total luminosities, above (shifted vertically from the data for visibility). It is intended that you be convinced that these are good fits.

Near Peak Luminosity

Monitoring of SN 1987A has continued on a reasonably regular basis and with a range of resolutions at optical, infrared, and ultraviolet wavelengths.[48] So has the modeling. The first feature to be noticed is the broad hump in the light curve at days 10 to 130, which is the equivalent of the plateau phase in normal type II's—that is, it represents energy released as a recombination wave moves inward. The amount of energy radiated during this phase tells us that the hydrogen-rich envelope had not been eroded to less than about 3 M_O by mass loss before the explosion. The observed broadness and smoothness require that the recombining envelope have its helium and/or carbon–oxygen layers rather extensively mixed with the hydrogen.[49] We will return repeatedly to the topic of mixing in the progenitor and the ejecta (enhanced by the reverse shock propagating into the ejecta[50]).

The second important point is that total brightness would have declined catastrophically after day 40 (the dashed line in Figure 3) without an additional energy source. Such a source had, in any case, been eagerly awaited by proponents both of nucleosynthesis and of pulsar formation in type II SNe. The light curves shown in Figures 2 and 3 have adopted 0.07 to 0.08 M_O of Ni^{56} as the additional source. Weaver and Woosley had earlier postulated such a contribution in exponentially tailed SN IIs,[51] and Uomoto and Kirshner reported exponential decline in $H\alpha$ emission line intensities.[52] It is worth noting (Figure 5) that the light curve of 1987A eventually coincides with typical ones. We perhaps saw evidence for the first breakout of Ni^{56} decay energy at about day 25, in the form of kinks in the color curves[53] and in the line excitation mechanism.[54] The actual energy source is gamma rays and positrons, released as Ni^{56} decays to Co^{56} (half-life 6 days) and in turn to Fe^{56} (half-life 77 days). In order for us to have seen material heated this way as early as the light curve indicated, even Ni^{56} (freshly synthe-

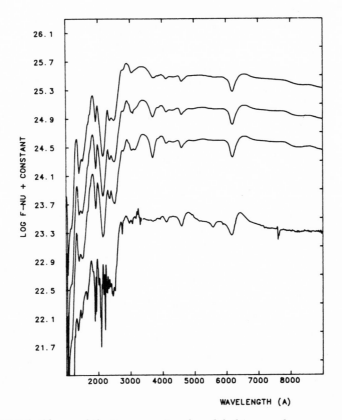

FIGURE 4. *Observed (bottom curve) and modeled (upper three curves) optical and ultraviolet spectra of 1987A, two days after turn-on. The models adopt a density varying as r^{-11}, with 1.5×10^{-14} g/cm^3 at the position corresponding to a velocity of 20,000 km/sec (about the largest hydrogen line velocity seen). The three curves from top to bottom have total luminosities of 4.9, 3.9, and 3.2×10^{41} ergs/sec, bracketing the observed value. Several features, including the extreme UV deficit, are well modeled. (Slightly recalibrated from two papers by J. C. Wheeler, R. P. Harkness, and Z. Barkat: in IAU Colloquium No. 108, ed. K. Nomoto [1988], and in Supernova 1987A in the Large Magellanic Cloud, ed. M. Kafatos and A. G. Michalitsianos [Cambridge: Cambridge University Press, 1988], p. 264. Courtesy of J. Craig Wheeler.)*

sized at the base of the ejecta) must quickly have distributed itself through the envelope. Another, and perhaps better, way to describe this situation is to say that some portions of the envelope became optically thin very quickly (owing to filamentation or fragmentation), so that the

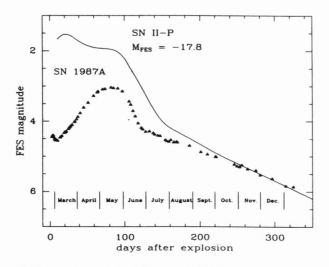

FIGURE 5. *Comparison of the light curve of 1987A (determined with the optical monitor on the IUE satellite) with the average of normal plateau-type SN II's. After about 200 days, when the light curve is largely dominated by Ni^{56} decay, 1987A ceases to be atypical. Courtesy of Nino Panagia.*

effective photosphere included a portion of material from deep layers of the star.

A third property of the light curve is that, once the Ni^{56} energy source has been included, then any pulsar contribution must be less than about 10^{40} ergs/sec, setting a severe constraint $B^2 P^{-4} \leq 2.6 \times 10^{-4}$, where B is magnetic field in units of 10^{12} gauss and P is the rotation period in milliseconds.[55] Light curves in which only a pulsar contributed would obviously be much less restrictive, but they have not yet been modeled in similar detail.[56]

Finally, we note another piece of evidence from this period for mixing in the progenitor. Strong barium and strontium lines[57] indicated the presence of s-processed material in the envelope (by no means uncommon in red giants and supergiants). A moderate excess (e.g., factor 2.5) of the full range of s-process nuclides (Sc, Ti, Ba, V, Cr, Sr) improves the fit of synthetic spectra to the continuum as well as to the lines.[58] Both the synthesis of these nuclei and their transport to the stellar surface require convection to have mixed several layers extensively.

Decline and Fall

The fall from peak luminosity joined at about day 130 onto an exponential tail. Since astronomers plot everything in log–log coordinates, this phase looks like a straight line in Figures 2, 3, and 5 and is often called the linear part of the light curve. Depending on precisely how observed wavelength bands are dereddened and summed,[59] this linear phase has an *e*-folding time of 103 to 115 days, in good accord with the 111-day *e*-folding life of Co^{56}. The summing is tricky because lines comprise more than half the total flux by day 350, and (a) available photometry misses some important ones (like Paschen alpha) and (b) one wants to include only the flux currently driven from the central energy source and not light echoes and the like. For instance, what looked for a while like an infrared excess due to heated dust[60] turns out to be line and band emission (especially CO at 4.8 μ) and powered from the center.[61] In fact, so far, we have no evidence for any emission from dust heated by the event at any phase.[62] Another infrared line at 17.93 μ[63] implies, first, that we are seeing a significant amount of iron made in the progenitor star and, second, that the radioactive mantle has been mixed with the hydrogen envelope (because the line extends blueward to velocities larger than the 2000 km/sec hydrogen minimum). The implications of some other infrared lines are addressed later.

The ultraviolet emission from 1987A has also continued to do interesting things. After the very rapid early drop, UV flux began to recover in May 1987.[64] The emission is largely concentrated in lines, which were still brightening in February 1988.[65] The gas responsible is considerably enriched in nitrogen (N/C = 8 ± 4; N/O = 1.6 ± 0.8). The lines are narrow (FWHM ≈ 30km/sec), meaning that the gas is not part of the expanding ejecta. The most probable source is material shed by the progenitor when it was a red supergiant, compressed into a shell by the later fast blue giant wind,[66] and ionized by the UV burst which marked shock breakout on 23 February.[67] The continued increase in line flux indicates that the emitting material is about 10^{18} cm out; thus the star must have become blue again 10,000 to 20,000 years ago. A prediction of this picture of the emitting gas is that strong soft x-ray emission should turn on in about 10 years, when the outermost ejecta reach the shell.[68]

Several other predictions are possible. First, the IR to UV flux should begin to drop below a linear extrapolation of Figures 2, 3, and 5 when the ejecta become optically thin to gamma rays. This effect has probably been seen, the deficit amounting to about 8% at day 385, according to data from SAAO (South African Astronomical Observatory).[69] This drop is not fully supported by CTIO (Cerro Tololo Inter-American Observatory) data.[70] Notice in the figures that the two observatories have never precisely agreed, the SAAO points being brighter at all phases.

Next, as the energizing flux continues to decline and leak out more directly, the object should experience an onset of strong infrared line cooling, in which the luminosity drops below 10^{39} ergs/sec[71] and is energized by the less powerful, but longer-lived nuclide Co^{57} (half-life 272 days).[72]

Finally, if core collapse left a neutron star, we ought eventually to see it, either as an energizing pulsar that prevents the decline just described[73] or at least in the form of thermal x rays from its hot surface. These should remain above the detection limits of the x-ray satellites ROSAT and AXAF for about 100 years[74] though one hopes for launches somewhat sooner.

Nucleosynthesis, X Rays, and Gamma Rays

Invoking 0.07 to 0.08 $M_\odot$ of Ni^{56} to explain the optical light curve of SN 1987A led immediately to the prediction that photons from the decaying, excited, product nuclides should become directly visible as the optical depth of the expanding ejecta dropped. The expectation[75] was that gamma-ray lines at 0.847 MeV and 1.238 MeV should turn up sometime in 1988, with a three-month warning signal consisting of a hard-x-ray continuum of nuclear decay photons degraded by multiple Compton scatterings.[76]

Even as the preprints were being mailed out, the x rays had already turned on in August 1987 and were being recorded by experiments carried by the Ginga and MIR satellites.[77] These x rays were much earlier than expected, considerably fainter than expected, and showed no signs of the rapid rise and exponential fall anticipated. If they were the Compton-scattered gamma rays, then the early onset could only mean that some Ni and Co were in zones of quite small optical depth, yet another piece of evidence for mixing[78] of a more thorough kind than that required for gamma-ray heating to contribute to the light curve at and beyond maximum light.[79]

How did these very deep layers find their way out so quickly? The energy released in the decays is comparable with the kinetic energy density of the local gas. Thus the surroundings are heated, expand, and push against the denser, cooler overlying layers. Such situations are Rayleigh–Taylor unstable and should lead to bubbles or plumes of Ni–Co-rich material popping out.[80] The early turn-on then marked the epoch when the first nickel bubbles reached moderate optical depth. The faintness and absence of rapid rise and fall mean that a few percent of Co^{56} was uncovered initially and that more was being revealed over the next year or more. Under these circumstances, gamma rays should also have started coming straight out earlier than expected.[81]

Once again, observations overtook predictions, with the announcement by Matz *et al.* that their NaI spectrometer on the Solar Maximum Mission (SMM) satellite had been seeing a modest excess of gamma rays peaked around 847 keV (though with very poor energy resolution) also since August 1987.[82] As with the x rays, the photons were early and faint (representing a few percent of the total Co^{56} implied by the optical light curve), and the flux was neither rising nor falling rapidly.

Given our anxiety to see the nuclear-decay gamma-ray lines, and the rather poor angular and energetic resolution of the SMM detector, skepticism was briefly permissible. But three balloon flights have confirmed an 847-keV flux near 10^{-3} $\gamma cm^{-2}s^{-1}$ and seen the 1.238-MeV line as well. The detectors were a low-resolution spectrometer launched from Alice Springs, Australia, in November[83] and two higher-resolution germanium spectrometers, launched from Alice Springs in October[84] and from Williams Field, Antarctica, in January.[85] These also confirm that the gamma rays are coming from the direction of 1987A.

The relative steadiness of the flux over several months has permitted (or required, depending on your point of view) refinements of the mixing models.[86] Sandie *et al.* also report a continuum flux between 50 and 200 keV, presumably representing another portion of the Compton-degraded spectrum seen by Ginga and MIR.[87] According to 1988.3 models, the gamma-ray flux should change rather little over the next year or so before entering into exponential decline once there is no more buried Co to be revealed.

Additional balloon flights occurred or were planned through the spring and summer of 1988. According to rumor,[88] these were expected to confirm the gradually increasing fraction of expected Co being seen (from 1.3% at first detection[89] to 10% in March) and the preliminary report[90] that the lines were (a) double peaked, (b) redshifted by 1000 km/sec or so more than expected from the +270-km/sec velocity of the LMC,[91] and (c) absorbed by 20 to 30 g/cm^2 each for the 0.847-, 1.238-, and 2.60-MeV lines (when one requires that their de-absorbed flux ratios match their branching ratios). The line splitting is at least qualitatively consistent with a remnant whose center has been cleared out by a reverse shock. The redshifting appears perhaps also in other spectral regions, and we return to it later. The absorption should decline with time and provide a test of the basic mixing model.

Much puzzlement resulted from the report that the hard-x-ray flux was varying suddenly and erratically.[92] It can, however, be argued that the Compton-degraded flux is, in fact, steady or slowly rising,[93] while a softer, thermal component changes both intensity and temperature in complicated ways.[94] This at least shoves all the rotten eggs into one basket, since no aspect of the soft x-ray component is very well understood. The proximity to SN 1987A of other bright x-ray sources, including the variable LMC X-1, has somewhat complicated the determination of fluxes and variability.

Finally, optical depths are much smaller in the infrared, permitting us to probe the entire remnant for nucleosynthesis products. Preliminary data have appeared at 8 to 13 μ from the European Southern Observatory[95] and extending to longer wavelengths from the Kuiper Airborne Observatory.[96] Line identifications were initially disputed, but it now seems at least probable (a) that a line near 10.5 μ is due to Co II; (b) that it faded by about a factor of 2 between November and March, corresponding to the residual Co^{56} declining from 0.044 to 0.0023 M_O as expected; (c) that another possible Co II line near 18.8 μ disappeared between November and March; and (d) that the stable product iron made its appearance in a 17.9-μ line, attributable to Fe II, whose width is large enough to suggest partial mixing into the hydrogen-rich envelope.

Asymmetries and Extended Structures

Many different kinds of data tell us that SN 1987 is not and never was a point or spherically symmetric source. The assorted data do not, however, add up to demonstrate any single asphericity or extension. Some items, in fact, seem to be mutually inconsistent. Thus, the ordering of topics is largely arbitrary, and the following paragraphs can be read in any order (or not read, as you prefer, though in that case the order matters less).

Small-scale lumpiness is suggested by the structure which began to appear in the $H\alpha$ emission line in October.[97] The ragged appearance of known supernova remnants (and of nova remnants, which are much younger) indicates that filamentation and fragmentation are common phenomena. It is interesting that the breaking up begins very early. A critical question is when filamentation progresses far enough to let us see a central pulsar or pulsar-driven nebula, and is this by any chance within the first year.[98]

At least two observations indicate that something interesting and asymmetric (on the scale of the total ejecta) was going on as the light curve began to rise and (perhaps) the first radiogenic flux appeared at 25 to 40 days. First, the optical polarization changed in intensity, position angle, and wavelength dependence.[99] Several interpretations have been offered, including an overall spheroidal structure with axial ratio about 0.7[100] and a fast-moving jet, presumably driven by Rayleigh–Taylor instabilities of hot underlying gas. The latter would imply a causal as well as a temporal connection with radiogenic breakout. Couch[101] and Cropper et al.[102] have suggested and Barrett[103] denied that the dominant polarization angle (after correction for interstellar contributions) is close to the 194° position angle of possible larger structure.

Second, between days 25 and 60, the hydrogen line profiles, both optical and infrared, showed triple peaks, with extra emission at 4000 km/sec on either side of the central maximum.[104] This transient emission (most thoroughly studied by the Bochum group, though the Kavalur data were published first) must have come from some asymmetric structure—a rotating ring or ellipsoid or oppositely directed jets—that formed or was uncovered at a fairly definite epoch and then dispersed or became optically thin. Lucy suggests excess emission energized by asymmetric first emergence of radiogenic x rays.[105] If this is the right interpretation, then there may be some connection with the red shifted infrared and gamma-ray lines[106] which probe the present distribution and motion of material that was originally in the innermost ejecta. The event perhaps needs some descriptive name; referee Robert P. Kirshner suggested "April split." Given the visual impact of the data shown by Hanuschik, Theim, and Dachs[107] and Phillips,[108] I am inclined toward "April banana split."

These spectroscopic and polarimetric data seem to pertain to things happening within the main expanding remnant. In contrast, an assortment of results from speckle interferometry and direct imaging pertain to larger volumes. Much the most famous of these is an apparent companion, seen in March and April by two groups[109] and not heard from since.[110] The image was 2 to 4 magnitudes fainter than the main supernova image (brightest in the red) and about 0."06 away in position angle 194° (all numbers having fairly large error bars attached). Called the mystery spot, son of supernova, and even worse things, the companion requires information, energy, or matter to have traveled at $\gtrsim 0.4 \ c$ if it was energized from the central event. This is not necessarily impossible, and might even be expected for a jet driven by a central neutron star[111] or merging pair,[112] which must then have hit previously existing material. The large luminosity of the companion presents problems for any model in which the triggering energy goes out spherically, as one would expect for the initial ultraviolet burst.[113]

One of the groups reporting the companion has also found the supernova itself to be resolved from late March 1987 onward[114] at several wavelengths. The measurements through their $H\alpha$ filter give radii of 0".008 in June 1987, 0".020 in February and March 1988, and 0".027 in April 1988, corresponding to average outflow of 4000 to 6000 km/sec, a reasonably representative velocity for the hydrogen-rich ejecta. The later images are moderately elliptical ($a/b \sim 1.2$). Strangely, the simultaneously measured sizes at blue continuum wavelengths are all as large as, or larger than, the $H\alpha$ images: 0".011 in April 1987, 0".018 in June 1987, and 0".026 in April 1988. This is much harder to interpret, and does not seem to correspond to any velocity or feature seen on other ways.

An L-band speckle observation, reported by Chalabaev et al.,[115] found still larger-scale structure in the form of spots or a ring providing 3 to 4 percent of the flux and 0".4 across in August 1987. This requires $v \gtrsim 0.4 \ c$ and so ought, presumably, to be interpreted as some sort of light echo (or as a last appearance by son of supernova, if it was the tip of a relativistic jet).

Such echoes are expected—in the infrared if dust is heated and reradiates[116] or in visible light if dust and gas merely reflect and scatter. The latter effect has been seen[117] in the form of two arcs or rings, about 30″ and 50″ from the supernova (in early 1988) and 5″ to 10″ wide. The inner one, at least, is moving out at about 1".8/month ($v \approx 19c$[118]) as was predicted from the geometry when they were first seen.[119] In early 1988, the arc spectra resembled that of the supernova at maximum light (April–May 1987), not that of the contemporaneous supernova. One could hardly ask for clearer evidence of the light echo phenomenon, first seen in Nova Persei 1901.[120] The echo is somewhat fainter and faster moving than predicted by Schaefer,[121] who thought it would remain at binocular visibility for some years, rather than requiring a coronagraph.

Residual Puzzles

Reference has already been made to several unexpected and perhaps unexplained aspects of 1987A. These include (a) the hows and whys and wherefores of the extensive and rather fine-tuned mixing required to account simultaneously for the optical light curve, the hard x rays, gamma rays, and infrared Co II lines if they are all results of the same decaying Co^{56} and (b) the short-lived speckle companion and unexpectedly large disk size at continuum wavelengths. This section addresses two more incompletely solved problems: the origin of the soft x-ray flux and the absence of evidence for a central pulsar.

One previous type II supernova, 1980K, was seen as a soft x-ray source by Einstein in its dying days.[122] The progenitor presumably had a standard red supergiant structure (since it displayed a typical SN II light curve) and had shed a dense wind late in life, thus accounting for both the x rays and strong radio emission[123] as coming from the shocked region where ejecta encountered wind material.[124] Such x rays were not expected from 1987A, given its blue progenitor and faint, brief radio emission, indicative of only tenuous circumstellar material. Behavior of the narrow UV lines mentioned above suggests that denser stuff is some 10^{18} cm out and should not be shocked for several years.

Nevertheless, the Ginga satellite saw x rays below 10 keV, beginning in August at the same time as the harder ones. Eight months of monitoring[125] revealed complex, and seemingly correlated, variability in the hard and soft channels. Tanaka[126] suggests, however, a deconvolution into two components, first the degraded nuclear gamma rays, with constant flux between days 190 and 390, adsorbed by about 10^{25} atoms/cm^2 or 5 g/cm^2 (assuming cosmic abundances, which may well be wrong) and, second, thermal bremsstrahlung, whose flux and temperature varied between 8×10^{36} ergs/sec at $kT \approx 4$–10 keV, and 10^{38} ergs/sec at $kT \approx 60$ keV. No absorption is evident in their energy band, but there must be a sharp cutoff below 2 or 3 keV to avoid conflict with simultaneous rocket upper limits[127] of about 1.5×10^{36} ergs/sec at 0.2–2 keV. The thermal nature of the emission is demonstrated by the presence (at least at high flux levels) of a 6.7 to 6.9 keV iron line. The measured energy means that the line is coming from iron atoms with only one or two electrons and is not fluorescence of neutral iron excited by initially nonthermal radiation.

The temporal history of the soft x-ray component is complex but can perhaps be summarized[128] as a baseline low, cool flux with a couple of moderate flares in September and November 1987 and a strong, hot one in January 1988. The rise time was about 10 days, and the return to baseline took about 25 days but with a 30-percent drop occurring in one or two days about 23 January. Peak flux reached 10^{38} ergs/sec, and the total energy in the event was at least 10^{44} ergs.

The January flare was, in a sense, predicted by Hillebrandt, Höflich, Schmidt, and Truran,[129] who said that soft x rays should turn on when the outermost ejecta hit a nearby cloud they had postulated to account for son of supernova. Detailed models along these lines are not, however, terribly satisfactory.[130] First, a normal thermal spectrum would extrapolate to a violation of the rocket upper limits at 0.2–2 keV. Second, we have not seen the expected associated radio emission. If 1987A produced the same ratio of 6-cm flux to 4-keV flux that 1980K did, then the baseline level would correspond to a 0.4-Jy source and the January flare to 5 Jy. But the limits (like the flux briefly seen) are at mJy levels. Masai[131] suggests that these objections are less severe if the collision is indeed with an isolated cloud rather than with a uniform circumstellar shell. But, third, the rapid fading on 23 January requires either a cloud with $n_H \geq 10^{11}$ cm^{-3}, if the time scale is set by radiative cooling, or expansion at v $\geq 10^9$ cm/sec, if cooling is adiabatic. Neither seems terribly likely.

Bandiera *et al.*[132] have suggested as an alternative that the entire x-ray spectrum is being radiated by a central, pulsar-driven nebula, whose radiation we see through a rapidly varying screen of fragmented ejecta, accounting for the changes in both flux and spectrum. Such a pulsar and nebula will inevitably contribute flux at other wavelengths as well, and more detailed modeling is required to decide whether it can also fit with the observed light curve, the independent (gamma-ray) evidence for radiogenic energy input, and so forth.

We are left with a puzzle: If this is not the pulsar, then where is it, and how and when can we expect to see it? The expected luminosity is

$$L = \frac{32\pi^4 B^2 R^6}{3c^3 P^4} \quad \text{erg/sec,}$$

where B, R, and P are surface dipole field, radius, and rotation period, all in cgs units. This is anything from 4×10^{44} ergs/sec on down.

The observations give only upper limits, mostly much lower than the maximum possible. TeV gamma rays are expected from the highest-energy particles accelerated in the magnetosphere but have not so far risen above 5×10^{38} ergs/sec,[133] considerably less than the predictions.[134] There is at most very marginal evidence for a TeV flux of about 3×10^{38} ergs/sec at the time of the January soft x-ray flare.[135] Once the radiogenic components in IR to UV, hard x rays, and MeV gamma rays are allowed for, limits of 10^{39-40} ergs/sec from the pulsar obtain at those wavelengths.

Apparently, then, the current rotation period (for $B = 3 \times 10^{12}$ gauss) is longer than 10 to 15 milliseconds, or the dipole field strength is less than $(1-3) \times 10^{10}$ gauss (for $P = 1$ msec), or collapse on 23 February 1987 continued past neutron star densities. Statistics of young pulsars suggest that the combination of rapid rotation and strong dipole field is, in fact, rare.[136]

Even if B^2P^{-4} is currently small, the relict neutron star could be producing up to 2×10^{38} ergs/sec in x rays fueled by accretion. In the absence of exotic cooling mechanisms like pion condensates or strange quark matter, thermal x rays should persist until satellites sensitive enough to see them to have been launched.[137]

To decide among the alternatives—pulsar, pulsar-driven nebula, inert neutron star with or without accretion, or black hole—we will have to wait either for emission from the relic to dominate or for the ejecta to become optically thin at some wavelength where emission is unambiguously expected. When this happens depends somewhat on composition and dust formation but mostly on the degree of filamentation. For 10 M_O of ejecta, moving at an average speed V_{4000} (in units of 4000 km/sec) for a time t years, the overburden is $10V_{4000}^{-2} t^{-2}$ g/cm^2 for a thin spherical shell, or $30V_{4000}^{-2} t^{-2}$ g/cm^2 for a uniform density sphere. This does not mean that we will never detect a central object without filamentation—on a clear day you look up through about 1000 g/cm^2, and 1987A is already nearly transparent at optical and infrared wavelengths. But it does mean that filamentation decides whether the time scale for visibility will be 1, 10, or 100 years at the more readily absorbed and scattered radio and x-ray wavelengths. If asked to bet, I would say that the time scale will turn out to be just about that on which most astronomers lose interest in a particular problem or object.

Update (1991)

Predictably, both the fluxes coming from SN 1987A and our perceptions of what they mean have continued to evolve. The gamma-ray lines and hard x rays faded, as expected, over several half-lives of Co^{56}, and the object has become much fainter at all wavelengths with the gradual loss of radioactive power input. Surprisingly, perhaps, no firm evidence for a pulsar had surfaced even four years after the explosion, though one false alarm came and went, putting a limit of about 10^{38} ergs/sec on pulsar luminosity at any wavelength. The light echo was resolved in a Hubble Space Telescope image in early 1991. This image, combined with the light curve of the echo from the International Ultraviolet Explorer satellite, has led to a very precise distance for SN 1987A near the expected value of 50 kpc. Meanwhile, theorists have used the properties of 1987A to fine-tune their models of supernovae in general, so that one of its long-term effects on astrophysics is likely to be a better understanding of which elements are made in which kinds of stars at different stages in galactic evolution.

So You Want to Find
a Supernova?

Although our basic understanding of supernovae has been stable for more than a decade, an assortment of large and small puzzles remains. Some of these can be expected to yield to statistical information from carefully designed searches; others seem to require detailed follow-up of selected events; and a few may still be awaiting a fundamentally new idea.

Many of us learned in childhood a guide to expository writing that said, "I keep six faithful serving men. They taught me all I knew. Their names are what and where and when and how and why and who." This presentation is organized very roughly along the lines of those six possible continuations of the rhetorical question, So you want to find a supernova? More complete treatments of the status of supernova research can be found in discussions by Trimble, Petschek, Woosley, and Wheeler *et al.*[1] The most up-to-date catalogue is that of Barbon *et al.*;[2] this contains all confirmed events through 1988. More recent ones are virtually all reported in the IAU Circulars.

How, When, and Who:
Supernova Discovery Rates and Successful Searches

Early supernova discoveries were serendipitous ones of naked-eye objects in our own galaxy. They were distinguished from novae and other transient phenomena only after the fact and not with complete certainty either of the supernova classification or of the type, I, II, etc. Telescopic discoveries began with S And (in M31) in 1885, and photographic recording of images and spectra commenced shortly thereafter.

Table 1 indicates numbers of extragalactic supernovae discovered per quinquennium since 1895, according to the list of confirmed events given by Barbon *et al.*[3] and the IAU Circulars for 1989–90. The two sudden increases in 1935–39 and 1950–54 reflect the initiation of deliberate searches, by Fritz Zwicky, using first the 18-inch Schmidt at Palomar and, later, the Sky Survey plates from the 48-inch Schmidt. A sharp drop occurs at Zwicky's death in 1974. Rarely has the number of discoveries exceeded the 26 letters of the alphabet, necessitating designations like 1954aa and 1954ab. 1990 was the third year in which this happened.

Between 1986 and 1990, about 20 different organizations and individuals have discovered one or more supernovae. These are listed in Table 2. Notice that well over half the events are still being found on photographic plates exposed on Schmidt telescopes and scanned by experienced human observers. Visual searches (especially that of Robert Evans in Australia) and the Berkeley automated search have also contributed significantly. Two special-purpose projects, the Distant Danish and Las Campanas, were intended to locate type Ia supernovae in rich clusters (at redshifts near 0.3 and nearby respectively) so that sets of representative light curves could be assembled for comparison with more distant SN Ia's observed by HST. In each case, fewer events were found than expected, and not all were type Ia.

The most productive searches probe rather different distance ranges, as reflected in the apparent magnitudes of the supernovae they find: 13^m to 14^m in the visual ones, 16^m to 17^m at UCB, 18^m to 19^m at CERGA

TABLE 1.   **Numbers of SNe Found per Quinquennium**

1895–99	2	1945–49	6
1900–04	2	1950–54	64
1905–09	2	1955–59	44
1910–14	2	1960–64	96
1915–19	3	1965–69	86
1920–24	5	1970–74	92
1925–29	7	1975–79	57
1930–34	1	1980–84	102
1935–39	18	1985–89	85
1940–44	8		

TABLE 2.   Succesful Supernova Searches, 1986A–1990Q

Automated		Visual	
UCB	11	Evans	8 ½
		Other	1 ½
Schmidt Patrol[a]			
POSS II	18	Other photographic	
Berne	6 ½[b]	ESO	2
Konkoly	2	Abastumani	1 ½
Asiago	4	Las Campanas	2
Sternberg	2 ½	AAO	1
CERGA/OCA	23 ½	Other (serendipitous)	
Kiso	1	VLA	1
Distant Danish	2	spectroscopic	1
Asteroid Patrol	1	UV CCD	1
UK Schmidt	1		

[a] The first five of these descend from projects initiated or inspired by Zwicky.
[b] Half events indicate nearly simultaneous independent discoveries.

(now incorporated with the former Observatoire de Nice as Observatoire de la Côte d'Azur), and 15^m to 19^m in the second Palomar Observatory Sky Survey.

Clearly, anyone contemplating new supernova searches should give some thought both to the properties of existing ones and to the specific scientific problems to be addressed. Only quite bright events are likely to receive any significant follow-up in the form of light curves and spectrograms, unless provision is made for acquiring these as part of the planned project. Sadly, an end seems already to have come to the golden era initiated by SN 1987A in which, for about two years, every supernova discovered was accorded at least one spectrogram for type determination (very often by A. V. Filippenko, who has probably not received the recognition due him for this service). On the other hand, samples large enough to determine SN rates in various kinds of galaxies can be achieved in reasonable times only by scanning for rather faint, distant events.

What, Why, and Where:
Supernova Conundrums and Contributions

Unanswered Questions

For present purposes, these questions can be separated into questions
where statistical information from well-designed searches can help directly, ones where it can help at best indirectly by telling us something
about SN environments and parent populations, and ones where some
new ideas seem to be needed. Unfortunately, the most basic problems
seem to belong to the latter categories.

In type II events, the energy derives from collapse of the burnt-out
iron (etc.) core, whose bounce sends a shock wave outward through the
stellar envelope. But most recent, careful calculations find that the shock
stalls somewhere en route and does not actually eject the outer layers.[4]
This prompt shock, a delayed shock, and other possible "pushers"[5] are
currently in competition as mechanisms to deposit sufficient energy in
the envelope, but at the moment we do not know what to look for in light
curves, spectra, or whatever to distinguish the possibilities. Arguably, it
would be of some help to have limits on progenitor masses from star
clusters and HII regions around the exploding stars, but this is possible
only for very nearby events.

For classical type Ia supernovae, in contrast, the physics of energy release is thought to be well understood—explosive burning of carbon and
oxygen (etc.) to iron (etc.) in degenerate dwarfs. The "best buy" model
still seems to be the merger of a close white dwarf binary whose total
mass exceeds the Chandrasekhar limit,[6] but the known binary white
dwarf population includes no pairs both massive enough to explode and
compact enough to merge in a Hubble time, despite several serious
hunting expeditions.[7] Searches with relatively brief follow-up might
plausibly help here by further constraining the parent population in
terms of its distribution perpendicular to the disks of edge-on spirals,
prevalence in E galaxies with and without evidence for recent star formation, etc.

In contrast to these two fundamental unsolved problems, there are
some subsidiary issues to which well-designed searches can contribute
more directly. One of these is the nature of type Ib SNe (no hydrogen
in their spectra, but fainter than Ia's by about 1^m5 and found only in
spirals). It has not yet been fully settled whether they come from
stripped stars whose progenitors were, on average, more (or less) mas-

sive than those of type IIs or from binaries whose parent properties can be expected to be intermediate between Ia's and IIs.[8] Other such issues are the extent to which Ib's should be further subdivided on the basis of the late-time behavior[9] and the extent to which II-L's and II-P's (linear and plateaued light curves) constitute separate populations versus a continuum.

Finally, there are probably still at least factor-of-two uncertainties in the absolute rates of SNe (per century per unit galactic luminosity) as a function of (a) SN type, (b) galaxy type, and (c) whatever else matters. A recent episode in this saga is the recognition that the "superproductive" Sc galaxies that have been seen to have three or more SNe in this century[10] are virtually all seen face-on.[11] This strongly suggests that they are telling us the right (high) rate, and that observational selection against discovering supernovae in edge-on galaxies is more severe than generally allowed for. I will return shortly to the difficult case of the rate of faint core-collapse events.

Interactions with the Rest of the Universe

These topics also include some where systematic searches are of immediate relevance and others where they are not. Supernovae are generally held responsible (perhaps via their remnants) for accelerating cosmic rays and for heating and stirring up the interstellar medium, creating fountains, a low-density hot phase, and so forth. The input required—or allowed—per event obviously scales with the SN rate assumed and, as things presently stand, may be awkwardly high for cosmic rays[12] and awkwardly low for other things.[13]

The same proportionality obtains for input of new heavy elements as a component in models of galactic chemical evolution. Currently those models have enough other free parameters to be able to accommodate any reasonable SN rate.[14] Searches sufficiently extensive to be able to establish SN rates as a function of parent galaxy metallicity would be valuable (but exceedingly difficult) in this context.

In contrast, searches must lead to precise light curves and spectra if they are to assist with using supernovae as distance indicators for measuring H_0 and q_0,[15] distinguishing cosmological red shifts from tired light[16] or metaphysical red shifts,[17] and probing presupernova mass loss from the effects of circumstellar material on light curves.[18] The connection between searches and supernovae as triggers of star formation,[19] pulsars as probes of the dense matter equation of state,[20] and other topics in applied supernovarology is still more remote.

When and What: The Rate of Core-Collapse Events and Existence of Subluminous SN IIs

The initially surprising properties of SN 1987A served to remind the community that not all supernovae reach "standard" peak brightness, so that reported SN II rates can set only lower limits to actual rates of core collapses in massive stars. Somewhat similar remarks can be made about faint SN I's, but the issue is less pressing because, for instance, the amount of nucleosynthesis expected from a Ia event is directly proportional to its luminosity. An accurate estimate of rates of core collapse is also important to those constructing neutrino or gravitational radiation detectors to study them.

The faintness of type II events in low metallicity environments was, in fact, predicted,[21] and, as the accompanying figure indicates, 1987A was not the first anomalously dim core-collapse SN, assuming, as is now usual, that the Crab Nebula was made by a type II and Cas A by a type Ib, which may or may not have been seen by Flamsteed.[22] As the Crab and Cas A illustrate, faint core-collapse events can eject anything from zero to lots of fresh heavy elements. SN 1885 (S And) was not particularly faint at peak light, but it did rise and fall remarkably quickly and would have been missed in a more distant galaxy by many standard searches.

Issues on which there seems still to be some residual disagreement include (a) What is the rate of "normal" SN IIs in bright spiral galaxies? (b) How many core collapses do we expect in the Milky Way? (c) What Milky Way rate is implied by the known historical SNe? and (d) How much evidence is there for sub-luminous SN IIs elsewhere?

Tammann[23] found 1.23 and 1.45 h^2 SN II per century per $10^{10} L_o^B$ in Sab-b and Sbc-d galaxies respectively. The searches carried out at Asiago and by Evans[24] imply lower values near 1.1 (same units) for the average of Shapley–Ames galaxies. This rises to 1.3–1.4 for all core-collapse SNe if the Ib rate is added in.

The Milky Way falls close to the line between Sb and Sbc and close also to an average Shapley–Ames galaxy with $L = 2 \times 10^{10} L_o^B$, leading us to expect a rate of relatively luminous core-collapse supernovae of 2.7 to 3.7 per century on the two scales. An independent estimate is possible from the (vaguely) known number of massive stars in our galaxy. Ratnatunga and van den Bergh[25] concluded that there should be only 1.0 (+1.8, −0.5) to 2.2 (+2.7, −1.6) core collapses per century, assuming a lower limit of 5 or 8 M_O for the Main Sequence masses of the progenitors. These numbers do not seem to allow much margin for subluminous

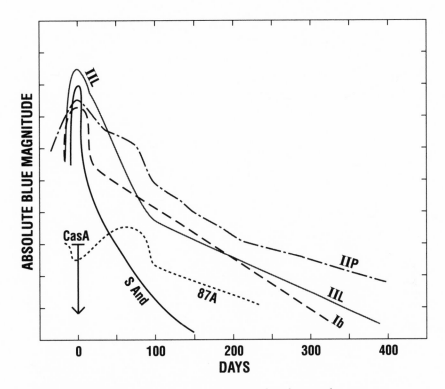

Impressionistic view of light curves of normal and anomalous core-collapse supernovae.

core collapses. Wood and Churchwell,[26] on the other hand, predict 1.84 to 4.0 core collapses per century, for a limiting mass of 8 M_O and still more for a lower mass cut.

What does the historical record say? Van den Bergh[27] interprets the 7 or so supernovae recorded between 1006 and 1685 as including 3 to 4 core-collapse ones within 3 kpc of us, implying a galaxy-wide rate of 2.1 to 2.8 per century. He regards this as anomalously high and calls attention to the presence of several spiral arms within the fiducial volume (but it would be difficult to draw a 3-kpc circle anywhere in a spiral disk without taking in a few arms). My own slightly discordant bookkeeping says that[28] there have been about 10 SNe recorded over the past 2000 years in the nearest 1/7 of the Milky Way. These are (with types in parentheses where I feel anything at all can be said) 185, 386, 393, 1006 (I), 1054 (II), 1181 (II), 1408 (II), 1572 (I), 1604 (I), Cas A (Ib). The implied core-collapse SN rate is 1.4 to 2.0 per century, and, once again, there does not seem to be much slack to be taken up by subluminous events. A firm upper limit of 1.5 core-collapses per *year*[29] is set by nondetection of neutrino bursts in the IMB proton decay experiment.

Finally, what evidence is there for faint type II supernovae among those seen outside the Milky Way? Based on the Evans search, van den Bergh and McClure believe that events like 1987A make up at most 0 to 30 percent of SN IIs in Shapley–Ames galaxies.[30] They suggest that this is the case because low-metallicity populations are rare in large spirals. Ron Mayle and James Wilson[31] conclude, however, that stars of initial mass 8 to 10 M_O should produce faint SNe (triggered by collapse of ONeMg cores) even for normal population I metallicity, because very little Ni^{56} will be made and the exploding stars will have very little hydrogen-rich envelope left (so that not much ionization energy gets deposited there for later reradiation).

The observational picture really depends on just how faint you mean by "subluminous." A very interesting culling of the catalogues by Miller and Branch[32] assigns peak B magnitudes (for $H = 75$) to 46 SN IIs from 1909A to 1989L. The mean and median are both just shy of $M_B = -17$, after allowing for absorption in the Milky Way but not in the parent galaxies. There is also some correlation of SN brightness with galaxy brightness, opposite in sense to what would be produced by absorption dominating the distribution of peak brightnesses. Only 1973R and 1982F were as faint in B as 1987A (peak $= -14.3$), but any reasonable correction for observational selection effects leads one to suspect that rather faint type II supernovae may be rather common.

The definitive search to resolve completely the faint SN II issue would probably have to go as faint as the brightness of ordinary novae ($M \approx -11$) and so to $m \approx 20$ even in nearby galaxies. This is not a realistic project for an automated telescope—or, at the moment, any other kind.

Conclusion

The last word belongs to Simon Newcomb (said to have been Walt Whitman's "learn'd astronomer"): "More than a score of such objects [temporary or new stars] are known to have appeared, many of them before the making of accurate observations, and the conclusion is probable that many have appeared without recognition."[33]

Remnants of
Historical Supernovae
WITH DAVID H. CLARK

The Historical Supernovae

The enormous stellar explosions called supernovae are sufficiently rare that none has definitely been spotted in our own Milky Way galaxy since the invention of the telescope. Thus students of the subject must either confine their attention to enormously distant extragalactic events and galactic remnants of unknown age or else turn to the writings of pretelescopic civilizations in search of dateable and locatable galactic supernovae. The events thus identified are the historical supernovae and number about eight.

From the earliest times of which we have any record, humans have watched the sky and attempted to interpret what they saw. Sometimes, their motives have been ones we share—keeping the calendar in step with the sun and moon, or deciding when to plant the corn. More often, perhaps, they had in mind what we would now call astrology—attempting to predict or even control terrestrial events on the basis of celestial configurations. As a result, they did not usually record precisely the information modern astronomers would like to have in the form we would like. So what one really means by "historical supernovae" is that you have to be something of a historian to interpret the data.

Not all civilizations have contributed equally to our knowledge. In particular early Indian literature has gone so far essentially unexamined for this purpose and may well harbor a wealth of reports of new stars as well as comets, eclipses, planetary conjunctions, and other astronomical phenomena. On the other hand, the most thoroughly searched literature of all, that of classical Greece and Rome, contains no certain records of any new stars (novae or supernovae) at all. The medieval European chroniclers reported the very bright supernova of 1006, and Renaissance scholars provided enough details of Tycho's (1572) and Kepler's (1604) supernovae to permit the reconstruction of approximate light curves.[1] Babylonian and Arabic records have been only partially examined, the former yielding no convincing new stars, and the latter a number of accounts of SN 1006 and a single reference to SN 1054.

The most productive investigations have been those of the Chinese (and to a lesser extent Japanese and Korean) astronomical literature. From the time of the Han dynasty (202 B.C.E.–220 C.E.) onwards, an astronomical bureau formed a subdepartment within the Ministry of State Sacrifices. Its two main functions were those alluded to above—maintenance of an accurate calendar and observation and interpretation of celestial portents. In addition, the records have been rather thoroughly examined by people knowledgeable in the languages and historical periods concerned.[2] Even here there are pitfalls. A great burning of books in 213 B.C.E. wiped out most of the history (and many of the historians) of what had gone before. Much of the provincial historical literature has yet to be examined for astronomical purposes (though work continues). And, finally, one must worry about both the reliability and the completeness of the available documents.

Reliability can be checked by means of solar system events whose times and directions in the sky we can now calculate. It seems to be quite good: the eclipses recorded are usually real eclipses, and the planetary positions recorded are accurate presentations of the phenomena. For instance, the calculated positions of Venus at times as long ago as the second century show that observations described in the Chinese records could have been made only on the day claimed, not two or three days earlier or later. Nevertheless, occasional grains of salt from an expert cellar are in order. For instance, many of the dynasties had particular colors closely associated with them, red for Han and T'ang, blue for Sui and Ch'ing, and a golden yellow for Shang and Sung. Reports of new stars in the reigning emperor's own color are therefore suspect (including the often-quoted description of the 1054 event as "yellow and favorable to the emperor").

Completeness is more difficult to assess, since we need to ask precisely whether absence of evidence is evidence of absence. An indirect argument comes from the temporal distribution of events recorded through a dynasty. Solar eclipses, needed for calendric purposes, were reported at a roughly constant and correct rate of three or four per decade throughout the Chin dynasty (*c.* 260–520 C.E.). But the situation for astrologically important phenomena is different. These phenomena include solar halos, sunspots, daylight sightings of Venus, and assorted conjunctions and occultations of planets with the moon, with each other, or with asterisms. Almost none of these turn up during the early "honeymoon" portion of the dynasty, while toward the end, as discontent increased, the omen rate increased to more than 50 per decade. The five candidate supernovae all come from the latter part of the dynasty, suggesting that there may be significant incompleteness in the early years. Other selection effects apply to other historical periods, and some of them can undoubtedly be mitigated by more careful examination of the provincial records. The implication is that, if an event is reported, it probably occurred, but that no strong conclusions can be drawn from the nonreporting of an event.

Several classes of temporary "stars" appear in the Chinese records. *Hui-hsing* ("broom stars" or "sweeping stars") are normally tailed comets; *po-hsing* ("rayed stars" or "bushy stars") are normally tailless comets; *k'o-hsing* ("guest stars" or "visiting stars") ought to include the novae and supernovae. But there are exceptions. Some *k'o-hsing* moved across the sky and so must have been comets. A very bright point source might have been perceived as tailed or bushy. And the 1006 event was called a *chou-po* ("Earl of Chou") star, apparently to emphasize its exceptional brilliance and auspicious nature.

We ask first of a candidate supernova that the event have remained visible for at least a month or two and that no motion have been recorded. These criteria yield the list of 19 events given in Table 1, though the 1408 star was reported only once, and neither duration nor position are well specified.

Some further elimination is possible. Supernovae in spiral galaxies are largely confined to their disks.[3] Thus any supernovae at galactic latitude b larger than about 25° would have had to be within about a kiloparsec of us and would have looked exceedingly bright. This means that the 5 B.C.E. and 61, 64, 70, 247, and 837 C.E. events were almost certainly comets or novae, though it is not always possible to decide which.[21] 1592A may have been Mira Ceti, and the other two events of that year complete fabrications. Only Korean sightings have been found. The remaining objects are the historical supernovae, listed in Table 2. Each has a remnant of roughly the same age certainly or probably associated with it.

Another handle on the completeness of the Chinese records comes from asking about young supernova remnants for which no initiating event has been identified. There are three—RCW 103 and MSH 11-54,

TABLE 1.   New Stars of Long Duration Recorded in China (C),
Japan (J), the Arab world (A), Europe (E), and Korea (K)

DATE	CHINESE DESCRIPTION	DURATION	RECORDED IN	REMARKS
− 5	hui	70+ days	C	comet/nova
+ 61	k'o	70 days	C	comet? (motion?)
+ 64	k'o	75 days	C	comet? (motion?)
+ 70	k'o	48 days	C	$b = + 45°$; nova?
+ 185	k'o	20 months	C	supernova
+ 247	hui	156 days	C	comet? (motion?)
+ 369	k'o	5 months	C	position unknown
+ 386	k'o	3 months	C	supernova?
+ 393	k'o	8 months	C	supernova
+ 396	"star"	50+ days	C	$b = −25°$; nova
+ 402	k'o	2 months	C	comet? (motion?)
+ 837	k'o	75 days	C	$b = +75°$; nova
+1006	k'o	several years	A, C, E, J	supernova
+1054	k'o	22 months	A, C, J	supernova
+1181	k'o	185 days	C, J	supernova
+1408	"star"	?	C	Cyg X−1?
+1572	k'o	16 months	C, E, K	supernova
+1592A, B, C	k'o	15, 3, 4 months	K only	Mira; fabrication?
+1604	k'o	12 months	C, E, K	supernova

both probably less than 1000 years old but too far south to have been seen from China, and Cas A, whose explosion either was not seen by anybody in 1657 ± 3[4] or was unexpectedly faint in 1679.[5] The implication is that the completeness of the records may be rather better than crude plots of events per decade would suggest. If so, then we can put reasonable confidence in the galactic supernova rate as derived from the historical events (~1/30 years).[6]

If the sample of Table 2 is reasonably complete for events within 5 kpc and in our sector in galactocentric longitude, it also makes sense to ask how probable the clumps and gaps in the temporal distribution are. The answer[7] is more probable than you might have guessed. A long simulation of randomly firing supernovae in a thin-disk galaxy showed at least 10 percent of all millennia having distributions as

TABLE 2.   The Historical Supernovae

DATE	OBSERVER(S)	REMNANT
185	Chinese	RCW 86 = G 315.4 − 2.3
393	Chinese	CTB 37A/B = G 348.5 + 0.1 or G 348.7 + 0.3
1000 ± 200	(extreme south)	MSH 11 − 54
1006	Chinese, Japanese, Europeans, Arabs	PKS 1459 − 41
1054	Chinese, Japanese, Arabs	Crab Nebula*
1181	Chinese, Japanese	3C 58 = G 130.7 + 3.1
1408(?)	Chinese	CTB 80; Cyg X-1 (both doubtful)
1572	Chinese, Koreans, Europeans (Tycho)	3C 10 = G 120.1 + 1.4
1604	Chinese, Europeans (Kepler)	G 4.5 + 6.8
1679(?)	Flamsteed	Cas A

*Otherwise known as Messier 1, Strohmeier 40, JH 357, GC 1157, Milne 9, 4C 21.19, van den Bergh, Marscher, and Terzian 8, 3C 144, NGC 1952, Hall 050211, CM Tau, 3A 0532 + 219, 4U 0532 + 21, NP 0532, H Ø534 + 21, CGS 0531 + 219, CG 185 − 5, 2CG 184 − 05.

clumpy as that from 985 C.E. to the present. Nevertheless Table 2 leaves the impression that we are overdue for another naked-eye supernova.

There is a considerable literature for all the objects in Table 2. But we wish to focus here on the Crab Nebula, the first of the historical supernovae to be identified, and still the most thoroughly studied (though not necessarily the best understood).

The Crab Nebula

Table 3 summarizes the history of discovery and our understanding of the Crab Nebula, supernova, and neutron star. The names and dates after 1900 come largely from *Jahresbericht* and its successor, *Astronomy and Astrophysics Abstracts*, and complete references can be found there.

TABLE 3.   Capsule History of the Crab Nebula, Its Pulsar,
and Related Phenomena

DATE	EVENT
1054	Light reached earth (July 4) from explosion *c.* 5283 B.C.
1731	Crab Nebula discovered by John Bevis
1758	Catalogued as M1 (Messier)
1844	Named (Lord Rosse)
1854	Emission genuinely diffuse, not due to unresolved stars (Lassell)
1916	General relativity and the Schwarzschild solution. High-velocity emission lines in the Crab (Slipher, who also very nearly looked for optical polarization and would have probably found it, *cf.* Brecher 1985).
1920	S And (SN 1885) had $M_v \sim -15$; Crab nebula = remnant of 1054 event (Lundmark)
1921	Crab shows large proper motions, amounting to rapid expansion (Lampland) Division of the novae into two magnitude classes is "not impossible" (Curtis)
1928	Star near center of the Crab has large proper motion (van Maanen)
1930	Theory of relativistic, electron-degenerate stars (Chandrasekar)
1932	Discovery of the neutron (Chadwick)
1933	(Dec.) Supernovae defined; energy comes from neutron star formation; some of it goes into accelerating cosmic rays (Baade and Zwicky)
1934	First neutron star models (Tolman) First supernova search (Zwicky)
1935–58	Supernovae become respectable: papers by Zanstra, McCrea, Barnothy and Forro, Hubble, Payne, Whipple, Sawyer, Humason, Popper, Mayall
1939	Neutron star models with approximate maximum mass (Oppenheimer and Volkoff) Continued gravitational collapse to less than Schwarzschild radius (Oppenheimer and Snyder) Supernovae can be used as standard candles for cosmology (O. C. Wilson) Neutrino emission a possible trigger for core collapse in massive stars (Gamow and Schoenberg)
1940	Two types of supernovae (type I and type II) can be distinguished from their spectra (Minkowski)
1941–42	First supernova review papers Neutron stars as gravitational lenses (Zwicky) Supernova theories (Gamow, McVittie, Schatzman)
1942	South preceding star is the remnant of 1054 event (based partly on spectral data) (Baade and Minkowski)
1948	Crab Nebula identified as a radio source (Bolton and Stanley)
1950	Type I supernova light curves are powered by nuclear decays (Borst)
1952	Supernova remnants as a class are radio sources (Hanbury-Brown and Hazard)
1953	Optical synchrotron emission predicted in the Crab Nebula (Shklovskii)
1954	Optical synchrotron (polarization) verified in Crab (Dombrovsky)

continued on page 201

TABLE 3. *(continued)*

1956–57	Stellar implosions and Cf^{254} as supernova energy sources; photodisintegration of iron as trigger of core collapse; nucleosynthesis in stars, including e and r process in supernova events (Burbidge, Burbidge, Fowler, Hoyle, Cameron, Christy, Baade)
1957	First Ph.D. dissertation on the Crab Nebula (Woltjer) Detection of radio polarization in Crab (Mayer, McCullouch, and Sloanaker)
1960	Thermonuclear explosions as an energy source for supernovae (Hoyle and Fowler) Cosmic rays accelerated in supernova shocks (Colgate and M. Johnson)
1963	Identification of Crab x-ray source (Bowyer, Byram, Chubb, and Friedman)
1964	Calculations of x rays from cooling neutron star models (Morton; Chiu and Salpeter; Hayakawa and Matsuoka; Bahcall and Wolf; Tsuruta and Cameron . . .) X-ray source not compact (Bowyer *et al.*), requiring continuous input of relativistic electrons Detection of compact, low-frequency radio source near Crab center (Hewish and Okoye)
1965	First neutron star review paper (J. A. Wheeler) Magnetic neutron stars (Woltjer)
1966	Momentum transport by neutrinos as ejection mechanism for type II SN (Colgate and White) Neutron stars should be detectable from x rays emitted when they accrete gas from companions in close binaries (Zeldovich and Guseinov)
1967	Rotating neutron star models (Hartle) Rotating, magnetic neutron star as possible energy source for the Crab Nebula (Pacini) Sco X-1 modeled by accretion on neutron star in close binary system (Shklovskii) (Aug.-Nov.) Pulsars discovered (Bell and Hewish)
1968	(Apr.) Second Crab Nebula Ph.D. thesis (Trimble) (Oct.) Discovery of pulsar in/near Crab (Staelin and Reifenstein; Comella *et al.*) (Nov.) Pulsar NP 0552 slowing down (Richards and Comella) = (Dec.) continuous energy input, as required (Gunn and Ostriker; Golderich and Julian)
1969	South preceding star = NP 0532 (Cocke, Disney, and Taylor; Nather, Warner, and MacFarlane; Lynds, Maran, and Trumbo) Second derivative of pulsar period (x-ray data, May, Goldwire and Michel; optical data, June, Boyton *et al.*; radio data, July, Richards *et al.*) (June) First conference devoted exclusively to Crab Nebula (Flagstaff) NP 0532 shows timing glitches (Boynton, Groth, Partridge, and Wilkinson)
1970	Discovery of Crab Nebula jet (van den Bergh) Neutron star mass $\gtrsim 1\ M_\odot$ needed to power nebula (Trimble and Rees) First international Crab Nebula symposium (Manchester, IAU Symp. No. 46)
1971	Upper limits to gamma-ray flux constrain Crab magnetic-field strength to roughly the equipartition value (Fazio *et al.*) 3C 58 pointed out as the first Crablike SNR (Weiler and Seielstad)
1972	Pulsations in binary x-ray sources show that the compact object is a rotating neutron star (Gorenstein *et al.*)
1972–79	Optical and UV spectroscopy of Crab Nebula show He/H>1,(C + N + O)/(H + He) = low to normal (Davidson; Kirshner; Henry; Gull; Fesen; Chevalier)

continued on page 202

TABLE 3. *(continued)*

1974	Pulsars are, in general, high-velocity objects (Rickett; Lang; Manchester)
	Binary pulsar and gravitational radiation (Hulse and Taylor)
1982	Millisecond pulsars (Backer *et al.*)
	Crab infrared emission comes from dust which accounts for about 0.01 $M_\odot$ of metals and raises heavy element abundance to normal (Glaccum and Harper; IRAS)
	Progenitor models of 8 to 10 $M_\odot$ (Nomoto; Hillebrandt)
1984	Dynamical data on the jet (Fesen; Shull)
	Pulsar in the LMC, 0540–69, shows x-ray (Seward, Harnden, and Helfand) and optical (Middleditch and Pennypacker) pulses at $P = 50$ millisecs, with a slowing-down timescale of 1200 years and many other similarities to the Crab and its pulsar

Most of this history is still very close to us in time and in the interconnectedness of workers in the field.

Some of the events listed now seem trivial, though they were important at the time, for instance Lassell's conclusion that the nebular emission was genuinely diffuse and not attributable to unresolved stars. The Crab was the first object for which this was firmly established. Other items, like the relatively large proper motion of the Crab's central star (first measured by van Maanen in 1928) came to seem significant only long after the event. The reality of this large velocity and the extent to which it might be shared by the nebula were still being debated in 1969.[8] Subsequent work on other pulsars has made a 100 to 200 km/sec^{-1} velocity seem likely enough, without definitely establishing whether the proper interpretation is liberation from close binary systems, asymmetry in the supernova, or something more esoteric.

Others of the items have never been properly published, including Landau's first conception of neutron stars in 1932 (described by Rosenfeld[9]) and the infrared data collected by Glaccum *et al.* using the Kuiper airborne observatory.[10]

Two topics on which there has been considerable discussion deserve slightly fuller treatment than is given in Table 3. These are the existence and frequency of supernova remnants resembling the Crab Nebula and the properties and causes of the jet.

Crab-like remnants, or plerions, are defined in terms of their radio properties, and, although about a dozen are known in the Milky Way (and a few more in the Magellanic Clouds), existing radio surveys have been carried out in ways that discriminate strongly against identifying them.[11] X-ray searches for compact cores in known supernova remnants[12] are much more complete, but they pertain only to recognized SNRs. Thus, to the extent that plerions (because of their structure and flat spectral indices) have not yet been separated out from HII regions in the radio surveys, the x-ray information does not really constrain their numbers either. The arguments of Srinivasan *et al.*[13] and others on the rarity of formation of rapidly spinning, strongly magnetized pulsars may therefore be subject still to some modification.

The Large Magellanic Cloud SNR and pulsar 0540 –69.3 have recently been advanced as near-twins to the Crab Nebula and 0532. They are also very nearly 180° apart in the sky, which ought to be a comfort to believers in cyclic cosmologies! Table 4 compares the objects. The numbers come largely from talks given by F. D. Seward, S. P. Reynolds, J. Middleditch, R. H. Becker, and R. P. Kirshner at the 1984 October meeting, "The Crab Nebula and Related Supernova Remnants."[14] To first order, one gets the impression that the similar properties are those dominated by the pulsar (except for pulsed radio emission), whereas those dominated by the original explosion and its interaction with the surrounding medium are different. Perhaps one should only conclude that stars of quite different masses can leave similar pulsars.

Van den Bergh[15] first reported a luminous jet extending from the northeast nebular boundary. Later work established that it emits both line (thermal) and continuum (synchrotron) radiation and prompted countless models, invariably a bit contorted to match the undoubted fact that the axis of the jet does not point back to the pulsar position either now or in 1054. These models will have to be rethought in the light of the observations (a) that the jet is moving outward along its own axis at about 4000 km/sec;[16] (b) that it is simultaneously expanding cylindrically perpendicular to its axis at about 360 km/sec;[17] (c) that the local magnetic field is aligned with the jet;[18] and (d) that the jet composition is quite nearly that of the adjacent nebular filaments, including less-than-average helium enrichment.[19]

The length of the jet divided by its axial velocity and the width divided by the transverse velocity both give time scales of 600 years. One is left with the impression that something merely punched a hole at a random weak point on the nebular surface, and the local mix of thermal and relativistic gas shot out into a less confining medium. It is probable and relevant that this jet is merely the most conspicuous of a fairly numerous class because it is bright and nearly in the plane of the sky. The velocity data of Clark *et al.*[20] show high velocity material at several points southwest and northwest of the nebular center, suggesting that we would

**TABLE 4.   Comparison of SNR 0540 – 69.3 and Its Pulsar
with Crab Nebula and NP 0532**

PROPERTY	0540	CRAB
Pulse period	50 ms	30 ms
dP/dt	4.79×10^{-13} sec sec^{-1}	4.23×10^{-13} sec sec^{-1}
dE/dt	1.5×10^{38} ergs/sec	4.5×10^{38} ergs/sec
implied surface field	4×10^{12} gauss	3×10^{12} gauss
L_x (Total)	1×10^{37} ergs/sec	2×10^{37} ergs/sec
$L_x/\dot{E}$	0.05	0.05
x-ray size	= optical	1/4 optical
x-ray spectral index	0.8	1.0
x-ray pulsed fraction	23% (larger at larger photon energies)	4%
x-ray pulse shape	sinusoidal	sharp pulse; wider interpulse
age from dP/dt	1600 yr	1230 yr
optical radius	1 pc	2 pc
Doppler expansion vel.	1250 km/sec	2000 km/sec
expansion age	800 yr	844 yr
real age	800–1000 yr (?)	930 yr
L_{opt} (total)	2×10^{36} ergs/sec	3×10^{36} ergs/sec
$(B-V)$ color	0.85 ± 0.35	0.5
opt. pulsed fraction	0.6%	0.4%
L_{opt} (pulsed)	10^{34} ergs/sec	10^{34} ergs/sec
opt. pulse shape	sinusoidal	sharp pulse and interpulse
Radio flux (408 MHz)	<1Jy	1000 Jy; 1.6 Jy at LMC dist.
Radio spect. index	0.43	0.25
Radio size, structure	= optical; plerion	>optical; composite
radio pulsed flux	<1/2 mJy	6 JY; 10 mJy at LMC dist.
remnant composition	O/H high He/H normal	O/H normal He/H high
estimated progenitor	25–30 M$_\odot$	8–10 M$_\odot$

see jets there if we happened to be viewing the object from a different
angle.

FRUITS OF
MARRIAGE

Gravitational Radiation and the Binary Pulsar

Gravitational radiation not only exists, but it is apparently also doing exactly what general relativity says it ought to, at least in the binary pulsar PSR1913 + 16. This was the conclusion drawn by J. H. Taylor and J. M. Weisberg in 1982.[1] It came after more than six years of observing gradual changes in the eight-hour orbit of the pulsar and its invisible companion, probably also a neutron star.

The pulsar period itself (0.059 sec) acts as the known emitted frequency from whose Doppler shifts the orbit is determined as part of a total solution including up to 20 variables. These represent intrinsic properties of the pulsar, classical orbital elements, and relativistic corrections for the advance of the perihelion (about $4°$ per year compared with $43''$ per century for Mercury) and variable parts of the gravitational red shift and the transverse Doppler effect. The relativistic parameters together define a small set of allowed values for the masses of the two stars. These plus the orbit period can be plugged into a standard formula for gravitational radiation from orbiting point masses to yield a unique prediction of the rate at which the orbit should be decaying, $\dot{P} = -2.4 \times 10^{-12}$. The measured value, from the multivariable fit, is $-2.30 \pm 0.22 \times 10^{-12}$, in exact agreement to within the experimental errors.

The high precision and self-consistency of these results mean that they have interesting implications for three problems of current interest in astrophysics. First, if no important effects have been left out, the binary pulsar orbit provides as accurate a mass for a neutron star as has ever been determined, and the first one for an honest-to-goodness radio pulsar. The other measured masses all come from x-ray binaries, for which the energy source is not the rotation of the neutron star (as in "true" pulsars) but accretion of gas from a companion, an inherently messy process. The answer is 1.4 ± 0.1 M_O for each component. Existing data[2] are consistent with all known neutron stars having about this mass. It is just above the maximum stable mass (Chandrasekhar limit) for a star or stellar core supported by degenerate electron pressure and may reflect a fairly universal truth: you get a neutron star when an inert core builds up to the Chandrasekhar limit and so collapses.

Second, we are encouraged to ask more seriously than before how gravitational radiation will affect the evolution of other short-period binary systems, whose orbits are less clean than that of PSR1913 + 16, but which are astronomically at least as interesting. The classic case is that of the cataclysmic variables (CVs)[3]—novae and related systems consisting of a normal star transferring material to a white dwarf companion. Many kinds of binaries show such mass transfer. In some, the flow is driven from a more massive to a less massive star by Newtonian effects; in others, a star spills over because it is trying to expand and become a red giant.[4] But in some low-mass CVs, neither of these explanations works. The shedding star is the less massive and has not yet begun to expand. What drives the transfer? Gravitational radiation must gradually bring the two stars closer together. The surprise is that it does so at precisely the rate needed to produce the otherwise inexplicable mass transfer in these short-period, low-mass CVs.[5] In addition, beyond a certain point gravitational-radiation-driven mass transfer begins to move the stars apart again, and so may also explain some puzzling gaps in the distribution of orbit periods of CVs.

Third, the binary pulsar provides a test of the standard formula, first derived by Einstein,[6] for the gravitational radiation expected from a binary system. Over the years, many theorists have tackled the problem within the framework of general relativity. Some have confirmed the so-called quadrupole formula,[7] while others have found different results or, more often, concluded that a proper derivation has not yet been carried out and is currently beyond them, too.[8] The agreement to within 10 percent between prediction and observation for the binary pulsar strongly suggests that the quadrupole formula is very nearly right, at least for relatively weak-field cases (again, unless some important effect has been left out of the multi-variable analysis). This should help theoretical relativity both with this particular problem (it is always easier to do a calculation if you know the answer in advance) and with others by suggesting appropriate approximation methods. General relativity is not the only theory of gravitation currently under investigation. But most of the others (except some cases of the Brans-Dicke theory) predict orbit changes due to gravitational radiation that disagree with the binary pulsar result.[9] We cannot, however, quite rule them out yet, if only because the calculations leading to that conclusion have been done with methods analogous to those that lead to the quadrupole formula for general relativity, which may, just possibly, not be appropriate.

What can we expect from PSR1913 + 16 and gravitational radiation in the future? Observations are continuing. And the nature of the problem is such that precision increases roughly with the square of the observing time. Thus another five years could narrow the error bars on the relativity parameters from about 10 percent to about 1 percent, providing neutron star masses with that precision and testing agreement (or disagreement) to 1 percent between the quadrupole formula prediction and the observed value for the orbit decay due to gravitational radiation. Beyond this we probably cannot go. Fluctuations in the Newtonian gravitational potential of the galaxy (due to stars, clusters, gas clouds, spiral arms and all the rest) cause random, undecipherable accelerations of a test particle and so will put noise into the pulsar timing measurements at a level near 2×10^{-14} on time scales comparable with the effects being sought.

Finally, once a phenomenon has shown up indirectly in an astronomical context, it is natural to ask whether it can be seen directly in the laboratory. Detectors designed to search for gravitational radiation pulses have been in fairly continuous operation one place or another for about 15 years. Ferrari et al.[10] have reported a recent data set in which widely separated detectors were simultaneously excited by some sort of background and reference earlier results. Helium-cooled detectors of higher sensitivity are being developed or are in operation at Rome, Stanford, Maryland, Tokyo, and other places. The expected gravitational radiation flux at the earth from the binary pulsar is, however, only about 10^{-8} ergs/cm^2/sec, less than 10 percent that expected from the Crab Nebula pulsar and some 10 orders below the sensitivity of foreseeable ground-based equipment at the relevant frequencies. Indirect methods will, therefore, clearly continue to be important for a long while—at least for PSR1913 + 16.

Masada, Suicide, and Halakhah

". . . speaking slowly, clearly, and loud enough to be heard above the chorus."

—From the stage directions
for the cantata Masada,
by MARVIN DAVID LEVY[1]

The Halakhic attitude toward suicide derives not as one might expect from the sixth commandment of the Decalogue (Exodus 20:13; Deuteronomy 5:17)—"Thou shalt not murder"—but rather from the Noahidic injunction (Genesis 9:5): "And surely your blood of your lives I will require; at the hand of every beast will I require it; and at the hand of man, even at the hand of every man's brother will I require the life of man."[2] Concerning the first phrase of this, a certain Rabbi Eleazar[3] remarked that it meant that I (God) will require your blood if shed by the hands of yourselves[4] (i.e., for committing suicide). Rashi[5] says that the prohibition applies even if no actual blood is shed (for instance, suicide by strangling). In the same passage (Baba Kama 91b) a somewhat ambiguous discouragement of suicide is attributed to Rabbi Eleazar Hakkapar. Concerning the passage (Numbers 6:11) "And make atonement for him, for that he sinned regarding the soul,"[6] he said: "Regarding what soul did this Nazarite sin unless by having deprived himself of wine? Now can we not base on this an argument *a fortiori*? If a Nazarite who deprived himself only of wine is already called a sinner, how much the more so one who deprives oneself of all matters?"

Hence, we have arrived at a general principle that suicide is something for which God will demand an accounting, and thus it is prohibited. The next step is to see how this applies. That is, how do we distinguish a prohibited suicide from other kinds of deaths, and how should we react to a suicide? At this point, we are already being somewhat unfair to the Sicarii at Masada; the distinctions in the form as they have come down to us seem to have been made as a consequence of the events of the Bar Kochba rebellion and the Hadrianic persecutions,[7] some 60 years after the fall of Masada, whose defenders could not therefore be expected to know about them, while much of the Talmudic commentary on suicide is even later.

Let us, therefore, continue to follow the development of the law, and then return to how it might be applied to the events at Masada.[8] The two kinds of deaths from which we must distinguish suicide are accidents (including acts performed deliberately, but while of unsound mind) and martyrdom. The former is the easier to deal with (at least in the context of Masada). It is said (*Semahot*, chap. 2):[9]

Who is to be accounted a suicide? Not one who climbs to the top of a tree or to the top of a roof and falls to his death. Rather it is one who says, "Behold, I am going to climb to the top of the tree, or to the top of the roof, and then throw myself down to my death," and thereupon others see him climb to the top of the tree or to the top of the roof and fall to his death. Such a one is presumed to be a suicide, and for such a person no (funeral) rites whatsoever should be observed.

Death Under Duress

Thus, a suicide requires both explicit intent or declaration (which implies that the person is sufficiently sound of mind to be capable of intention) and witnesses. This is emphasized later in the tractate by the statement that no rites should be denied to a person (i.e., he is not to be accounted a suicide) if he is merely found hanging or impaled on a sword or if he falls into the sea, is swept away by a river, or is devoured by a wild beast. Furthermore, two incidents are recounted of children (both boys) who took their own lives after being threatened by their fathers. Rabbis Tarfon and Akiba ruled that no rites were to be denied them. From these cases, the codes (e.g., *Tur, Yoreh de'ah* 345:3) derive the principle that a child is never to be accounted a deliberate suicide, not having reached the full measure of intelligence.[10] The same passage also exempts one "who takes his life under duress, similar to the case of Saul" (whom we will meet again soon). This appears to include a large fraction of the suicides recounted in biblical and Rabbinic writings. A nonexhaustive list of these follows.

1. Saul's armor bearer, who fell on his sword when he saw that Saul was dead (I Samuel 31:5)

2. Ahitobel, an unsatisfactory counselor to King David, who strangled himself when he "saw that his counsel was not followed" (II Samuel 17:23)

3. Zimri, an Israeli officer, who burnt the king's house down around himself after being defeated by General Omri (I Kings 16:18)

4. Bagesh, a respected man of Jerusalem (*c.* 150 B.C.E.), who, when the soldiers of Nicanor came to arrest him, killed himself by a succession of unpleasant methods (II Maccabees 14:41–46)

5. Ptolemy, an advocate of the Judeans at the Syrian court, who poisoned himself when accused of treason (II Maccabees 10:12)

6. The nephew of Rabbi Jose ben Joezer, who, in remorse for having ridiculed his uncle's faith and plight as a martyr, subjected himself to the four methods of Jewish capital punishment (stoning, burning, decapitation, and strangulation) sequentially (*Genesis Rabbah* 65:22)

7. Three boatloads of adult captives from Jerusalem who drowned themselves out of fear of what might be done to them (*Lamentations Rabbah* 1:45)

8. The priests who cast themselves into the flames at the destruction of the first temple (*Taanit* 29a)

9. The pagan executioner who joined Hanina ben Teradyon in the flames after being promised a part in the world to come (*Avodah Zarah* 18a)

10. The servant of Judah ha-Nasi who killed himself on hearing of his master's death (*Ketubot* 103b)

11. The Hasmonean girl desired by Herod, who jumped from a roof to escape him (*Baba Batra* 3b)

12. The 400 boys and girls carried off for an "immoral purpose" who cast themselves into the sea (*Gittin* 57b)

13. Hiyya bar Ashi, who, after lying with his wife when she was disguised as a famous whore, fasted to death in repentance for the sin that he had intended to, though did not, in fact, commit (*Kiddushin* 81b)

14. A man who hanged himself in shame after inviting guests and finding he had nothing to serve them (*Hullin* 94a)

15. A mother and father who killed themselves after the father had killed their son in anger (*Derekh Eretz Rabbah*, chap. 9, 57b)

Although some of these suicides were declared deserving of the world to come by a heavenly voice (*bat kol*), they are nevertheless to be taken as exceptions and not as paradigms for our behavior.[11] But because suicide under duress does not count as willful self-destruction (*ma'abed atzmo l'daat*), no funeral rites are to be denied to such persons (*Tur, Yoreh De'ah* 345). In the case of a willful suicide, according to *Semahot*, "There may be no rending of clothes, no baring of shoulders, and no eulogizing for him. But people should line up for him and the mourners' blessing should be recited over him, out of respect for the living. The general rule is: The public should participate in whatsoever is done out of respect for the living; it should not participate in whatsoever is done out of respect for the dead." The Codes of Maimonides[12] and Karo[13] concur in this distinction, while that of Rabbi Jacob ben Asher[14] states that immediate relatives do mourn the deceased and rend their garments as a sign thereof. Nahmanides was still more lenient in his interpretation of what honors are due the deceased,[15] and the later authorities have tended to follow him, though often stopping short of an actual funeral oration.[16]

Martyrdom

The second distinction we must make is between suicide, which is forbidden, and martyrdom, which is required under certain circumstances. That is, we must strike a just balance between the Rabbinic principles of preservation of life (*pikuach nefesh*) and sanctification of the Divine Name (*kiddush hashem*). The most extreme view would never permit a Jew to die if there were any available alternative: "Has it not been taught: Rabbi Ishmael said: Whence do we know that if a man was bidden, 'Engage in idolatry and save your life,' that he should do so and not be slain? From the verse (Leviticus 18:5) Ye shall therefore keep my statutes and my judgments; which if a man do, he shall live in them. But not die by them."[17] Rabbi Saadia Gaon (*Emunot V'Deiot*, essay 10, chap. 11) and the *Tosafot* (*Gittin* 57b) subscribe to this view. The Sages, however, ruled otherwise, in the following fashion.

A midrash[18] continues the discussion by Rabbi Eleazar (*Baba Kama* 91b) of what is meant by the Genesis verse (9:5) in the following terms: "And surely (*ve'akh*) your blood of your lives will I require. This includes one who strangles himself. You might think that even one in the plight of Saul is meant: therefore we have *akh*. You might think, even one like

Hananiah, Mishael, and Azariah: therefore we have *akh.*" This cannot be understood at all without the useful footnote,[19] "It is a principle of exegesis that *akh* and *rak* (excepting, save that) are limiting particles." In other words, these are meant to constitute exceptions (presumably two separate classes of exception, since *akh* is repeated[20]) to the general prohibition of suicide. The class exemplified by Hananiah *et al.* is the easier to understand.[21] They are martyrs in the classic sense; that is, they deliberately allowed their lives to be taken by being thrust into the fiery furnace rather than commit the sin of idolatry. (The fact that God saved them alive out of the furnace is in this context taken to be incidental to the story, although it is the most interesting part from other points of view.) Rather precise standards for martyrdom were established by a majority vote of the Sages meeting at Lydda about 135 C.E.

"Rabbi Johanan said in the name of Rabbi Simeon ben Jehosadak: By a majority vote, it was resolved in the upper chambers of the house of Nithza in Lydda that in every (other) law of the Torah, if a man is commanded: 'Transgress and suffer not death' he may transgress and not suffer death, excepting idolatry, incest (which includes adultery), and murder."[22] The ruling to transgress and live is derived from the verse, "he shall live in them" (Leviticus 18:5) and the ruling to suffer death rather than commit idolatry from the verse "And thou shalt love the Lord thy God with all thy heart and with all thy soul and with all thy might" (Deuteronomy 6:5), which is interpreted[23] to mean that rather than worship any other in place of God (and thou shalt love the Lord thy God) one should give up one's life (with all thy soul) as well as one's wealth (with all thy might).

A continuation of the same passage extends compulsory martyrdom to two other cases. First,[24] when Rabbi Dimi came, he said that "transgress and suffer not death" was taught only if there is no royal decree (prohibiting the practice of Judaism), but if there is a royal decree, one must incur martyrdom rather than transgress even a minor precept. And according to Raba son of Rabbi Isaac, speaking in Rab's name, a minor precept includes even changing one's shoe strap.[25] Second, when Rabin came, he said in Rabbi Johanan's name that even if there is no royal decree, to transgress rather than suffer death is permitted only in private; in public one must be martyred rather than violate even a minor precept. "In public" is then defined to mean in the presence of ten Jews. This follows, says Rabbi Jannai, the brother of Rabbi Hiyya bar Abba, from the usages of *tokh* (among) and *eidah* (congregation) in the following passages: (1) "But I will be hallowed among (*be'tokh*) the children of Israel" (Leviticus 22:32), (2) "Separate yourselves from among (*mi'tokh*) this congregation" (Numbers 16:21), and (3) "How long shall I bear with this evil congregation" (Numbers 14:27). The last refers to the spies sent out by Moses to examine the land of Canaan. There were, in fact, twelve of them (one per tribe), but two were willing to follow God's commands and try to conquer the land and thus were not "evil," leaving ten Jews to constitute a "congregation." Since the ten are required to be men for the purpose of public recitation of prayers, presumably the ten witnesses that would make martyrdom "public" would also have to be men.[26]

This classic definition of martyrdom contains two firm requirements. First, one must "suffer death," that is, allow oneself to be put to death, as opposed to killing oneself. Second, it must be either for the sake of one of the three fundamental commandments or when Judaism has been proscribed by royal decree or in public. Hananiah *et al.* would clearly have met these requirements had they not been saved by divine intervention. Many of the traditional Jewish martyrs also are clearly included in the definition: another Eliezar and the seven sons of a woman (Hannah in some versions, Miriam in others) who allowed themselves to be killed under Antiochus IV (about 168 B.C.E.) rather than eat swine and commit other sins (but the mother's subsequent death belongs to the "under duress" class; II Maccabees; *Gittin* 57b); the ten rabbis[27] in the martyrology read on Yom Kippur (executed in about 135 C.E. in various ghastly ways for persisting in teaching Torah in public); those of the Rhenish and York martyrs during the first and second Crusades (1096 and 1190) who were killed as an alternative to baptism (but not those who killed themselves and their children in anticipation); more recently, those members of Jewish Councils under Nazi governments who chose to be executed themselves rather than select members of their community for the honor,[28] thus preferring to suffer death rather than commit the sin of murder.

The Case of King Saul

The Sicarii who died at Masada clearly do not fall within this definition of martyrdom, since they took their own or each others' lives. But then neither does Saul, who fell on his own sword (I Samuel 31:4) or requested a passing Amalekite to slay him (II Samuel 1:9); yet Saul is specifically exempted (*akh*) from those who are guilty of willful self-destruction and thus have no part in the world to come. He is, evidently, the prototype of a second class of exemptions, and we must inquire into what defines this second class and ask whether it is a class to which death is compulsory (like martyrdom), permitted, or merely forgiven after the fact. Here, then, is a list of the various classes of deaths of which Saul might be a prototype and the attitudes which one or another source says we ought to have toward them.

1. According to Nahmanides,[29] Saul was pursued (*nirdaf*) and therefore suicide was not prohibited to him (hence, presumably permitted). Since the principle of pursuit normally entitles one to kill (if all else fails) the pursuer, rather than the pursuee, this would be a difficult precedent to apply. The characterization of Saul as "pursued" does not appear in the Rabbinic text on which Nahmanides is commenting, or in other writings.

2. Saul, having an absolute foreknowledge of his own death, was doomed (*avud*) and suicide was therefore permitted to him (*mutar lo*) according to Nahmanides, or even preferred to the alternatives.[30] The Codes (*Tur, Shulhan Arukn*) do not describe suicide in the face of doom as permitted, but only as exempt from deprivation of burial rites. But the heart of the permission seems to lie in the certainty of death.[31] Thus, a person surrounded by the enemy, with a sword at his throat, must not take his own life but rather must rely on God's mercy. Saul's certainty derived from a visit from the prophet Samuel (who happened to be dead at the time), who said to Saul, "Tomorrow shalt thou and thy sons be with me" (i.e., also dead; I Samuel 28:19). This kind of certainty is not available any more; hence, on this interpretation, Saul's death cannot be the prototype of a class of permitted suicides.

3. Saul might reasonably have expected that, if he were captured, the Israelites would attempt to rescue him, and many might be killed. Under these circumstances, one suicide might be permitted to prevent many deaths,[32] even though, in this case, the deaths cannot have been anticipated with certainty. This is a possible prototype with modern applications, though not relevant to Masada.

4. Since Saul was the "anointed of the Lord," it would perhaps be a profanation of God's name if he had been killed in a shameful manner by the uncircumcized Philistines.[33] Thus, his suicide was permitted by the principle of *m'nyat hillul hashem* (avoidance of the profanation of God's name). The extension from a king especially chosen by God to a whole people or nation regarded as holy to God is obvious.[34] Rabbi Goren (while Chief Chaplain of the Israeli Armed Forces) argued further that suicide to prevent national disgrace is not only permitted but a *mitzvah*,[35] provided that death is certain in any case.

5. The traditional definition of mental illness must exclude Saul, who was surely not an imbecile (*shoteh*), but his fear of torture by the Philistines might well have been great enough to constitute a mental aberration which would justify us in forgiving him after the fact. This appears to be the view of the largest single group of authorities (if not an absolute majority).[36] In this view, Saul is the prototype of suicides who are forgiven after the fact and to whom "no rites are to be denied."

According to Ritzvah[37] Saul feared that the Philistines would force him to betray his faith. Many authorities[38] declare a preemptive suicide permissible in this case. Rabeinu Tam[39] goes beyond this and declares that if there is fear that the heathen will force one to transgress by means of torture impossible to endure, then suicide is imperative and a holy deed. In this interpretation, Saul is the prototype of a new kind of martyrdom distinct from that defined at Lydda in that death is self-inflicted rather than suffered and the transgression is merely anticipated rather than immediate.

The Events at Masada

The time has come to consider the specific happenings at Masada. Our only contemporary authority for these is Josephus,[40] who claimed to have talked with survivors and may also have had access to Roman records that have since disappeared. Much energy has been expended[41] in debating whether he was telling the truth about the events, the presence of Sicarii (as opposed to Zealots) at Masada, and the religious nature of the "fourth school's" belief in absolute liberty.[42] It will be assumed here that he was, in order to keep the legal issues reasonably clear. The three issues are, then, the suicides of the people at Masada (should they have done it? what should our attitude toward them be?), the method by which it was accomplished (they killed each other, but should each rather have killed himself?), and the behavior of the Jewish slaves who, according to legend at least, built the ramp that brought the Romans to the top of the fortress walls (should they have done it?).

The last of these seems the simplest. The ramp builders undoubtedly contributed materially to the defeat of the fortress. They were, therefore, voluntarily acting to produce either the enslavement or the massacre of the defenders. (That the defenders took another way out is not relevant at the moment.) To act so as to promote a murder is regarded as almost equivalent to murder itself.[43] Thus the decision at Lydda would have required the slaves to suffer death rather than build the ramp if they were certain that the Romans intended to kill the defenders.[44] But if they expected that the Romans would merely enslave the defeated Sicarii, then there does not seem to be any duty to die rather than transgress. Their own experience as Jews enslaved by the Romans and brought from Jerusalem to Masada can hardly have convinced them that the Romans always killed defeated enemies. Thus, even if the slaves are judged by standards that did not exist until many years later, their actions seem justified.

Concerning the method of death, there is a minority view that it is better to have another kill you. Hence, Rabbi Joshua Rosanis says that although one who commits suicide is punished from heaven, when he asks another to kill him his punishment is of a lesser degree. Thus, says Rabbi Rosanis, Saul asked others (his armor bearer and the Amalekite) to kill him to lessen his own punishment.[45] The trouble with this view is that it does not consider what happens to the person who does the killing for you. He is surely guilty of murder. Nor is your request any excuse.[46] King David was evidently of this opinion, as he had the Amalekite killed (II Samuel 1:14). One who kills himself does not fall under the category of "thou shalt not murder." He is subject only to the death penalty from heaven.[47] Thus, following the principles regarding *pikuach nefesh*, where we transgress that which is most lenient, most authorities have concluded that it is at least preferable for a person to kill himself and perhaps that this is the only permissible way.[48] Minors (boys under thirteen years of age, girls under twelve and a half) may constitute an exception, since they can hardly be expected to kill themselves. Thus a parent or other adult may perform the act for them under conditions where suicide is permissible.[49] Evidently, then, it would have been preferable for the adults on Masada each to have killed himself rather than choosing lots to kill each other.

Finally we come to the most difficult question—whether or not the suicides at Masada were permitted or even commanded by Jewish law at one or another of its stages of development. Should we regard the deaths at Masada as 960 martyrdoms, 960 cases of willful self-destruction, or something in between? How should the law be applied? The deaths cannot in any sense be regarded as accidents. The speech of Eleazar[50] ben Yair declared that the people intended to kill themselves, and then they did. Nor do they fall within the classic definition of martyrdom found in Sanhedrin 74a. Thus, if they are something other than willful suicides, it must be because they fall into the second class of exemptions defined in *Genesis Rabbah* (9:5) by King Saul. The first three interpretations of Saul's act (see "The Case of King Saul," pages 216–218) are not relevant, the first because it is not clear how to apply it, the second because the Sicarii could not have been certain of death at Roman hands, and the third because there was not anyone likely to try to rescue them. This leaves three possibilities: (4) a permitted suicide, *m'nyat hillul hashem*, (5) a forgivable suicide under duress, or (6) a martyrdom in the sense of Rabeinu Tam. No one seems to have applied precisely the Maharshal's justification of Saul's being allowed to commit suicide, which is our (4) to the Masada case. (Rabbi Goren[51] goes further and makes the deaths a *mitzvah*.) I should like, therefore, to propose this as my own small addition to the already excessively voluminous literature on the subject.

The next of the possible views, that of suicide under duress, proscribed but forgivable, seems to be supported by the majority of authorities.[52] Particularly interesting is the passage by Rabbi Yechiel Epstein[53] which suggests as a possible mitigating circumstance the "misconception that such a suicidal act was a *mitzvah*."

The last possibility, that the Sicarii at Masada were true martyrs, has been strongly espoused by Yadin and Spero.[54] They must, at best, have been martyrs in the extended sense of Rabeinu Tam, since they killed themselves rather than suffering death. In addition, we must find a commandment in which they can be described as preferring to die rather than transgress. That commandment is clearly not one of the three (to refrain from murder, incest, or idolatry) accepted at Lydda as grounds for martyrdom. Rather, the commandment must have been one to refrain from serving any master but God, corresponding to the great love of liberty which Josephus attributes to the "Fourth Philosophy" (that of the Sicarii). Such a commandment can be derived from "but if a servant shall say . . . I will not go out free, then shall his master bring him unto God . . . and his master shall bore his ear through with an awl, and he shall serve him forever" (Exodus 21:5–6). Rashi explained this by saying, "the ear that heard at Sinai—they are my servants and not servants to servants." Thus, evidently, a Jew has an *a priori* responsibility not to serve any master but God. And it seems not unlikely[55] that if the Sicarii had been voting at Lydda, the commandment not to serve would have been one of the cardinal ones for which the Jew was bidden to suffer death rather than transgress. Under such a law, those who died at Masada would surely have been martyrs, and our judgment of them should perhaps be tempered accordingly.

Conclusion

Willful self-destruction is forbidden and martyrdom required by Jewish law, while suicide under duress is forbidden but forgiven after the fact. The majority of authorities considering the deaths of the defenders of Masada have placed them in the third, forgivable, category, along with the death of King Saul. We have seen, however, that there is one line of thought within the mainstream of halakhah (point 4 in "The Case of King Saul," pages 215–217) which would call the deaths permitted (though not compulsory) because of *m'nyat hillul hashem*.

In addition, Masada fell long before the decision at Lydda restricted martyrdom to cases where the alternative was idolatry, murder, or sexual immorality, a decision which was not described as unanimous even then. Another group of Sages (especially if it had included some of the Sicarii) might well have decided that serving men instead of God was as much a sin as serving false gods. According to their gemara, those who died at Masada would have been true martyrs, at least in the extended sense that allows active as well as passive deaths.

Gravitational Radiation Detectors

T he story of the search for gravitational radiation really begins in the mid-1950s, when a small child in suburban Maryland developed the curious habit of banging his head repeatedly against the side of his crib in the dead of night. This gave his father, a physicist at the University of Maryland with a background in the technology of radio engineering and electronic countermeasures, many extra hours for solitary contemplation. He began reading and thinking about Albert Einstein's theory of gravity—general relativity—and how it might be brought into closer contact with modern laboratory physics.

The three experimental tests of general relativity known at that time explored only the small deviations from Isaac Newton's theory expected in the relatively weak and static gravitational field of the sun. The subject was even less well developed than was electromagnetism before Heinrich Hertz demonstrated the existence of radio waves in 1888. But Einstein's equations predicted that much larger deviations from Newtonian gravity would occur when compact, massive objects moved around rapidly, and that information about these rapidly changing, strong gravitational fields could move through space as a wave, bringing knowledge not available any other way.

Some scientists doubted the physical reality of these gravitational waves,* and standard reviews of the subject universally proclaimed that they would be forever undetectable. Joe Weber disagreed. And, by 1960 (when his son had gone on to more interesting activities than head bang-

---

*People who work in experimental relativity use the phrases "gravitational radiation" and "gravitational waves" more or less interchangeably. They also sometimes speak of "gravity waves," but this phrase was long ago adopted to describe certain purely classical oscillations of the earth's atmosphere and similar systems. It should be used for the relativistic waves only in contexts that do not invite confusion.

ing), he had designed and published a description of a device that could act as a gravity-wave antenna.

The detector was a large cylindrical bar of stiff aluminum alloy, which would be set into vibration when struck by a pulse of gravitational radiation. Onto it were cemented piezoelectric crystals that generate an electric field when squeezed or stretched. Thus vibrations of the cylinder would cause a current to flow in wires attached to the crystals. Amplifiers and receivers could then detect and process the signal.

Of course a great many things besides gravity waves could cause the cylinder to vibrate and current to flow in the wires Any interesting signal has to be seen against this background noise. Noise sources include the unavoidable electric and magnetic fields of normal laboratory equipment, power lines, and lightning, as well as the vibrations of the floor excited by moving people, passing traffic, and minor earth tremors.

Despite these difficulties, and with the help of several graduate students, technicians, and a post-doctoral researcher, Weber had an antenna built and working by 1965. Inside a vacuum chamber and hung from acoustic filters, it was isolated from light switches and heavy feet well enough for the receiver output to be dominated by the thermal energy of the aluminum bar itself. This fundamental noise source can be reduced only by cooling the antenna well below room temperature.

Weber quickly realized that the only hope of distinguishing signals due to gravity waves from all the other disturbances would be to run two antennas at the same time, some distance apart. He could then compare their outputs and look for instances when both were set into vibration. Unfortunately, both antennas will occasionally be noisy at the same time quite by chance. The rate at which such coincidences occur must be calculated or, better yet, measured before the experimenter can be sure that there is some signal above and beyond the background. A good way to measure the chance rate is to correlate the signal from one antenna with that from the other taken over a different time interval. Ideally this is done by someone who does not know in advance which data points represent simultaneous outputs and which shifted ones.

Coincidences above the expected chance rate started turning up in 1967 in the outputs of two antennas placed on opposite sides of the University of Maryland's College Park campus. This was a bit too close for comfort. But the Argonne National Laboratory, 1000 kilometers away in Chicago, generously offered a home for Weber's second bar in 1968, and the search was on.

Gravitational radiation leaped into scientific and even popular view with the 1969 announcement that the Argonne and Maryland antennas were showing statistically significant simultaneous excitations. Weber's 1970 report that these events tended to occur when the detectors were pointed toward the center of our galaxy suggested a possible astronomical source.

To understand why these announcements caused interest, excitement, consternation, and a flurry of detector construction in the physics community, we need to back off and look at the properties of gravitational radiation.

The Nature of Gravitational Radiation

Gravitational waves turn up in several different ways of thinking about physics and the universe. Some sound very esoteric indeed. For instance, the equations of general relativity with which Einstein unified geometry and gravitation have solutions that represent propagating changes in the curvature of space–time. This means that a triangle, whose angles add up to $180°$ in flat space, will oscillate when a gravity wave comes past. The sum of the angles alternates between more than $180°$ (positively curved space) and less than $180°$ (negatively curved space). Clearly you could look for this effect. Just place observers with lasers and mirrors at the corners of a giant triangle and ask them to keep track of the angles they see and report back to you from time to time.

Of course, there is a catch. The fabric of space–time is extraordinarily stiff. A gravity wave can carry a great deal of energy and still distort space only a tiny bit. More familiar waves have this property too. Think of twanging a rubber band and a tightly stretched piano wire in the same way; the stiff wire will wiggle much less. Well, space–time is more rigid than steel by a factor of something like 10 billion. Thus gravitational waves from even the most powerful sources will change the angles of your triangle by only about 10^{-34} degree! Because this effect is so small, none of the operating or proposed detectors make direct use of it. The geometrical way of thinking is, however, useful in calculating how much power can be expected from different sources and how sensitive particular detectors will be. Because both space and time are distorted, calculating the quantities actually measured in the laboratory can be tricky.

Another approach is the particle physics point of view, in which gravitational radiation is the wave aspect of a massless, chargeless, spin-two particle called the graviton. Quite remarkably, the people who think about this sort of thing for a living (they are called quantum field theorists) say that the properties of this particle require gravity to behave in just the way described by Einstein's general-relativistic equations. Once again, none of the existing or planned detectors look for individual gravitons. But this field theory way of looking at gravity is the best bet for some day unifying it with nature's other three forces into a single, "supersymmetric" theory.

Analogies with Light

Perhaps the most insightful way to think of gravitational waves is by analogy with light, x-rays, radio, and other familiar forms of electromagnetic waves. If charged particles (in the filament of a light bulb or a radio broadcast antenna, for example) are wiggled back and forth, they generate electromagnetic radiation that will move through space at the speed of light and wiggles other charged particles (in the retina of your eye or the radio receiver antenna) when it hits them. By the same token, oscillating masses should generate gravitational radiation that will move through space at its own characteristic speed and cause the oscillation of other masses when it encounters them. The sources and detectors described here can be thought of in these terms.

The speed of gravitational waves is nearly that of light—exactly so from a theoretical point of view if the graviton is truly massless, and experimentally so to within about five percent from the observed rate at which Mercury's orbit precesses.* In addition, a gravitational wave resembles an electromagnetic one in having a wavelength and a frequency whose product is the speed of propagation. In both cases, the wave's period is just the reciprocal of its frequency.

But there are also important differences between the two kinds of waves. Electrical charges come with both plus and minus signs, while all masses are positive. As a result, the simplest gravity wave represents motion more complicated than just an up-and-down wiggle. If this is forgotten, it is possible to orient a detector so it cannot see a particular source or,

---

*One of the three classical tests of general relativity is the rate of advance of the perihelion of Mercury. The predicted value is an algebraic expression that includes the speed of gravity. The observed value of the advance, 43 arc seconds per century (with an uncertainty of less than 10 percent), pins down the speed to be very close to c, the speed of light.

two detectors so they cannot possibly see the same source at the same time. The complicated shape of the oscillation means, in addition, that a pulsar or other object rotating with period P will emit most of its gravitational radiation in waves with period $\frac{1}{2} P$. And since the motion is relative to other particles, a detector must consist of an extended object (like Weber's bars) or at least two separated point masses (like the interferometers mentioned later).

The dominant difference, however, is that the force of gravity is devastatingly weaker than that of electromagnetism. The ratio is 10^{-40} for the simplest possible system, an electron–proton pair. One result is that most astronomical objects emit very little gravitational radiation—about one percent of the electromagnetic radiation from "good" sources like close binary stars and less than 10^{-20} of the electromagnetic power from "bad" sources like single stars. Another result is that even rather large detectors are not very sensitive. For instance, supernova explosions put enormous amounts of energy into pulses lasting about a second. One of these coming from the center of the galaxy in the form of gravitational waves would just barely show up in the output of existing multi-ton detectors. The same pulse in visible light (if there were no dust between us and the galactic center) would strike your eye with a brightness rather greater than that of the noonday sun.

Our understanding of gravitational radiation did not reach this level along a smooth path. Many physicists, notably Leopold Infeld and sometimes Einstein himself, argued that the wave solutions in general relativity did not represent real energy transport, making detection impossible even in principle. Definitive proof that the mass of a system decreases when it emits gravity waves came only in 1962 with work by Herman Bondi and Roger Penrose in England and Ezra T. (Teddy) Newman and Rainer Sachs in the United States. To this day, many astronomical sources remain far too complicated to study with exact calculations. Their gravitational radiation power must be determined using computerized numerical techniques or standard approximations.

Astronomical Sources

Every planet, star, galaxy, or cluster in the universe is a possible source of gravitational radiation as it moves in orbit or changes its shape. The earth–sun system, for instance, emits about 100 watts of gravity-wave power at a frequency of two cycles per year; our distance from the sun decreases about ten millionths of an inch per million years as a result. Strong sources of radiation at detectable frequencies are much rarer. The most promising candidates are close binary systems, supernovae or other stellar collapses, rapid pulsars, newly forming or evolving quasars and their ilk, and events in the very early universe. (These are ordered according to the level of our understanding of what to expect from them, from relatively high to very low.)

Einstein's first, simplest wave solution to the equations of general relativity was for an orbiting pair of point masses. The expression he derived for the power was correct, at least as a first-order approximation, and is still generally used. It was applied in the late 1950s and early 1960s to predict that a close pair of white dwarfs or neutron stars would eventually merge in a great burst of gravitational radiation at kilohertz frequencies. It was also used to show that some nova and dwarf-nova binaries could be spiraling together fast enough to account for the observed flow of gas between the stars.

The former effect motivated Weber's design for the first generation of gravity-wave detectors. But such mergers are probably rare, and we are not sure that any have ever occurred. The latter effect is now included in all standard models for cataclysmic binary evolution. But these binaries have so many other things going on at the same time that the effects of gravitational radiation on their orbits cannot be properly separated out.

More recently, Joseph Taylor and his colleagues at Princeton and the University of Massachusetts have found a much cleaner star pair, the famous binary pulsar. The pulsar moves in an eight-hour orbit with another neutron star, and Taylor's careful timing of the pulses reveals that the orbit's size and period are shrinking at just the rate predicted by Einstein's formula. For many astronomers, this discovery decisively established that gravitational radiation both exists and behaves as general relativity says it should.

The radiation from the binary pulsar has a dominant period of four hours, too low to be seen directly with even the next generation of antennas. But other similar systems undoubtedly exist and will eventually be studied.

Predicting the gravity-wave signature of a collapsing star is a bit trickier. Large spiral galaxies like the Milky Way host a type II supernova (in which a massive star's core collapses) about once a century. But nobody is sure what fraction of the available energy will come out in gravitational waves rather than neutrinos and other forms, or how long the pulse will last.

If supernovae release about half their energy in a quick burst of gravitational radiation, then galactic events will excite existing detectors. Nearby extragalactic events should show up with the next generation of instruments. If the pulses last more than a second or contain less than one percent of the energy, which is equally possible, then detection becomes much harder.

The resulting pulsars will be gravitational radiation sources only to the extent that their shapes are more complicated than smooth ellipsoids. The power is likely to be small. In addition, it comes out at the awkward frequency of 60 Hz (for the Crab Nebula pulsar), twice that of the pulsar rotation, where terrestrial laboratories are very noisy and detectors hard to isolate. At the moment we know only that gravitational radiation is less important than electromagnetic in slowing down the Crab and other pulsars. The late Hiromasa Hirakawa and his colleagues in Tokyo have set the best laboratory limit, using a detector tuned to the frequency expected from the Crab pulsar. They missed seeing it by a factor of at least 100 million.

Still more uncertain are estimates of the strengths and frequencies of gravitational waves radiated by collapses and collisions in the nuclei of quasars and active galaxies, by the processes that formed stars and galaxies in the distant past, and by events close in time to the Big Bang. Most of the energy is likely to be at the very low frequencies best studied from space rather than from the earth's surface. We know an upper limit to how much there can be: the gravity-wave energy at all frequencies put together cannot be so great as to dominate the total energy of the universe.

Most experimental limits are much less sensitive than this cosmological constraint. There are two exceptions. First, the very constant periods of some millisecond pulsars tell us that they are not being shaken around very much. The resulting limit on gravitational wave power at periods near a year is less than one percent of the cosmological limit. Second, and less widely known, is a suggestion from Martin Rees of Cambridge University that gravitational waves with periods of millions of years and wavelengths of millions of light-years—relics from the early universe—could be shaking the galaxies in clusters, giving the impression of large random speeds and, therefore, of excess (dark) matter. The required amount of radiation is about at the cosmological limit.

The history of astronomy suggests that when astronomers and physicists finally agree that gravitational radiation has been seen directly, the strongest sources will be things that nobody predicted in advance. Thus it is by no means unreasonable to record data with apparatus that "shouldn't" see anything. The first radio and x-ray telescopes shouldn't have seen anything either.

Astronomical and Space-Based Detectors

We have already noted the observable role of gravitational radiation in changing the orbits of close binaries and the limits on low-frequency waves that can be set by watching pulsars, galaxies, and the universe. Other branches of space science provide detectors for higher frequencies. Those so far tried include spacecraft, the earth, and the moon. An interplanetary spacecraft and the earth can act as the two separated masses of a detector. If they are wiggled enough by a passing gravitational wave, the frequency of the radio signals sent between them for tracking purposes will shift back and forth perceptibly. The greatest sensitivity is to frequencies like cycles per hour.

The tracking data for Mariners 6 and 7 (and, later, Pioneers 10 and 11) were examined for such effects after Weber's 1969 announcement. One possible Mariner glitch occurred at nearly the same instant as a laboratory event, but these data probably set only upper limits that are not as sensitive as the cosmological ones. Among the problems are that frequency shifts can be caused by interplanetary plasma and other things besides gravity waves, and that there are not always two spacecraft available at the same time to look for coincident events. Physicists at the Jet Propulsion Laboratory and the University of Pavia, Italy, had planned to surmount these difficulties by simultaneously tracking Galileo to Jupiter and Ulysses around the sun. Their plans are necessarily now on hold until these probes can be launched and reach suitably large distances from the earth. If the launches are too far apart, simultaneous tracking may never be possible.

The earth and moon are also possible gravity-wave antennas. Each can resonate in many different patterns having periods from 1 to 60 minutes. The signature of a gravitational wave is a vibration in those patterns, and only those patterns, that match the funny quadrupole shape predicted by Einstein. A gravimeter to monitor relevant earth vibrations has been recording data in Weber's lab at the University of Maryland almost continuously since 1963. In 1972, the Apollo 17 astronauts emplaced a lunar-surface gravimeter, provided by the Maryland group, to perform the same function there.

Most of the noise in the earth data comes from small, distant earthquakes. By the same token, the lunar data quickly revealed that "moonquakes" triggered by meteorite impacts are more common and longer lasting than expected. NASA stopped collecting lunar gravimeter data after just a few weeks, though the instrument is, of course, still on the moon. Again, only upper limits have so far come from these data, but gravimeters at quieter sites might probe below the cosmological limit.

Spacecraft tracking and earth–moon vibrations are sensitive to about the same frequencies of gravitational waves. It would make sense, therefore, to look for coincidences between them. Several sets of data covering the same periods of time show events that excited at least two of these "detectors." Since noise sources that affect only one can then be ignored, such a combination is much more sensitive than either alone. Back files of both tracking and gravimeter data exist on tape, and trying to correlate them is one of the many good ideas in science that has yet to be funded. The possibility of correlating with earth-based gravimeter data will make the Galileo and Ulysses tracking information valuable even if the two cannot be launched at about the same time.

Laboratory Detectors

Since astronomical sources of gravitational radiation were supposed to be weak and detectors insensitive, Weber's 1969 announcement of simultaneous signals in Chicago and College Park was something of a shock. Droves of theorists hastened to explain—or explain away—the events, to recalculate the power output of assorted sources, and to compute the sensitivity of hypothetical new detectors. At least fifteen experimenters made plans to build their own detectors, improving, they intended, on the original design. Eight or ten of these were eventually built, though in some cases they collected data for only a few days or weeks. None were precise copies of the original bars. Some used smaller masses of sapphire instead of aluminum cylinders; others did not control the bar temperature in the same way; still others extracted the signals from the antenna or processed them with very different techniques. Many of them operated in isolation rather than as coincidence experiments. And none of them saw anything.

Both the scientific literature and the popular press began reporting the negative results in 1971, sometimes in less than gentlemanly language. Fur and scurrilous adjectives flew to the extent that I shall always think

of one of our colleagues as the Gingham Dog.* One result of these skir-
mishes was a dramatic falloff in funding for gravity-wave experiments at
Maryland and elsewhere. Another was a sufficient muddying of the historical waters that it seems neither edifying nor perhaps even possible to
sort out who said and did what, or when, or why.

Room-temperature bar antennas have continued to operate at Maryland through thick and thin, sometimes seeing events, sometimes not.
The source, whether or not gravitational in nature, is not constant. In later years, groups operating antennas in Rome, Munich, and Tokyo have
reported positive results for coincidence experiments between their bars
or theirs and the Maryland ones. Some of these also reproduced the correlation with direction to the galactic center. Neither these reports nor
that of pulses recorded by a single low-temperature antenna at Stanford
in 1982 attracted much attention, though the latter had much the same
energies and time scales as implied by the Maryland data.

Before lowering the curtain on this medieval period in the history of
gravitational radiation, it is worth taking note of two prescient experiments. Robert L. Forward, who had been a graduate student at Maryland
in the 1960s, followed up on Weber's design from that period with an antenna using two separate masses. He collected data with such a device at
Hughes Research Laboratory in 1971. And the Maryland group in 1972
cooled a full-size cylinder down to only a few degrees above absolute
zero with liquid helium. Unfortunately, this system was so complex that
it hardly ever all worked at once, and the experimenters learned more
about cryogenics than about gravitation.

When the curtain rose a few years ago on the modern era, the people
who were still—or again—interested in building and operating gravity-
wave detectors presented a much more united front. Only two main kinds
of antennas seemed worth pursuing: multi-ton Weber-type bars of aluminum or niobium cooled to the lowest possible temperatures, and sepa-
rated-mass devices extending over several kilometers. Because the masses are not tied to each other and their separation is measured using the
interference of laser light, these latter detectors are often called free-mass
interferometers.

Institutions that have cooled bars to liquid-helium temperatures include Stanford, Louisiana State University, and the Universities of Rome,

---

*According to poet Eugene Field, "The Gingham Dog and the Calico Cat side by side on the table
sat." Some 30 lines later nothing is left but bits of gingham and calico, and "not a trace of cat or pup;
'tis said they ate each other up."

Western Australia, and Maryland. Current bar antennas, like the one currently operating at Rome, bear a striking resemblance to the primordial Maryland one.

Recording of data with these new instruments has so far been sporadic.* This reflects the difficulty of working at extremely low temperatures: every time something goes wrong, it takes three months to warm up your antenna, fix it, and cool it down again. Furthermore, standard conservative theory says that these detectors will have to be taken down to the still lower temperatures—less than 0.01 degree above absolute zero—provided by helium-3 cooling to have a good chance of sensing events regularly.

The main exponents of the interferometer systems are at the Max Planck Institute in Munich, Glasgow University, MIT, and Caltech. All have operated prototypes with masses separated by a few meters, and all are seeking money to build kilometer-size arrays. Whether these detectors are built above or below ground, the space between the masses must be kept as a vacuum to permit accurate measurements of their separation using lasers. The combination of vacuum equipment, electronics, people, and all the rest is an expensive one. A Caltech–MIT project comprising two 4-kilometer arrays at sites in California and New England carries a price tag of $60–80 million. The principal investigators, Ronald Drever and Rainer Weiss, are actively seeking the necessary sums and are hopeful of success within a few years.

The multi-mass detectors are intrinsically sensitive to gravitational waves over a range of frequencies, and the bars can be made so. Both work best between about 30 and 10,000 cycles per second (Hz). Luckily, this is still where theorists expect much of the power from supernovae and dying binaries to be. Measurements at frequencies much lower than 1 Hz, expected for radiation from active galaxies, young binaries, and the early universe, require spacecraft tracking or, eventually, multi-mass devices placed in space. One design carries the charming name "Skyhook." Because the several techniques complement each other, it makes sense to pursue all of them.

What can we expect in the future? My involvement with the search for gravitational radiation goes back to 1972, when I met and married Joe Weber, and we both confidently expect to be around to provide you with another progress report in 2002. By then, both millidegree bars and large

---

*Only three gravitational radiation antennas were operating at the time of supernova 1987A in the large Magellanic cloud. All were room-temperature bars—two at Maryland and one at Rome.

free-mass antennas will have been collecting data for a decade or so. Stellar and binary collapses will surely have been identified and studied, the gravitational waves coming from deep within their cores bringing us information that the light from their surfaces never could. But, undoubtedly, the most exciting results will have been things none of us predicted. And I fully expect that some bright-eyed junior of the class of 2003 will be in my office asking, "Gosh, Dr. Trimble, do you really remember back before the discovery of gravars?" Oh yes, I do indeed.

ASTRONOMERS OBSERVED

The Best Is Yet to Be

We are all more or less conditioned to believe that scientists are most productive very early in their careers. This has undoubtedly been true for many, including some, such as Newton and Einstein, whose contributions were truly revolutionary. But it is not the whole story, at least for outstanding (though nonrevolutionary) twentieth-century American astronomers. Helmut Abt (editor of *Astrophysical Journal*) has charted out[1] the publication histories of 115 people who received doctorates in astronomy in the United States between 1945 and 1960, and the citation histories of papers written at various ages by 22 very productive astronomers whose careers were largely complete before 1970. Both studies yield surprises.

The 1945–60 cohort, apart from moderate peaks associated with the award of doctorates and the "tenure crunch" 5 to 8 years later, have continued to produce research papers in refereed journals at a nearly constant rate over the next 25 years—despite an immediate postdoctoral loss of 9 percent of the group from the ranks of the published and an annual 1.5 percent loss thereafter. Thus, the most active members of the cohort show publication records that increase monotonically with time until, 20 to 25 years postdoctorally, the 13 most productive are accounting for 65 percent of the published pages. The members of this group have now reached career stages at which international scientific reputation and participation are likely to peak. (The only journals scanned were *Astrophysical Journal*, including *Letters* and *Supplements*, *Astronomical Journal*, and *Publications of the Astronomical Society of the Pacific*, the three American periodicals that publish original research papers over the full range of astronomy and astrophysics. Papers were normalized to 1000-word pages divided by the number of authors of the paper. Inclusion of non-American journals, conference proceedings, and review publications in the data base would probably have resulted in a still steeper rise with time in published pages for the more prolific authors.)

Contrary to other cherished prejudices, the high-productivity group (all are men) are not preferentially employed by full-time research institutions and are not much over-represented in the National Academy of Sciences, relative to other members of their age group. Several are widely known for their willingness to take on committee, conference, and administrative responsibilities; others are rarely seen outside their own offices. Abt concludes that productive astronomers are largely self-motivated and manage to function well regardless of environmental pressures.

Even less expected is the record of paper citations, which provides some measure of how influential individual authors have been at various times (though it is possible to be too influential: most papers on Hertzsprung–Russell diagrams do not cite either Hertzsprung or Russell). For this study, astronomers were selected as having published frequently in *Astrophysical Journal* before 1970; their work was therefore largely done in the United States, but most received their degrees before 1945 and there is little overlap with the previous group. Data came from *Science Citation Index* and so reflect papers from a very wide range of periodicals and books, for which the individuals investigated were senior or sole authors.

The first surprise is the sheer number of citations—9435 (from 1970 to 1979) to nearly 1000 papers (published from before 1910 to after 1970; median 1951) by 21 men and 1 woman (130 to 1400 citations each; mean 430). The second is that half of all citations are to papers published by authors 58 years old or older, and this drops only to about 53 when correction is made for the exponential fall (half-life 24 years) in citation rate as papers age past their peak influence, achieved 5 years after publication.[2]

Of the 22 authors, four wrote their most cited papers in their 70's, six in their 60's, four in their 50's, five in their 40's, one in his 30's, none in their 20's, and two were bimodal (30's + 50's and 30's + 60's). Many show a relative minimum in middle life, representing the war years when their attentions were wholly or partly directed away from astronomical research. At least four of the cohort continued to publish after 1970 and piled up a few dozen more citations to these late papers during 1980 and 1981, though in no case enough to affect the total statistics.[3]

What are we to make of all this? First, astronomers apparently live a long time (all 13 most productive of the 1945–60 cohort are still hard at it; 5 of the 22 earlier astronomers garnered 82 citations to papers written past age 80). Second, we can apparently expect to go on producing work that our colleagues will consider worth publishing and referencing for as long as we remain motivated to try. And, for most of us, if not our best work, at least our most cited work is probably still ahead of us.

Information Explosion

T he scientific information explosion needs no introduction. More journals are publishing more papers by more authors on ever-more-specialized topics than ever before. And, at least within the major American journals of astronomy and astrophysics, there are now also more words per paper than ever before.[1] The rise in mean paper length started shortly after World War II, since when it has been monotonic and more than linear with time, with the result that a typical 1980 paper is nearly three times wordier than typical 1910–40 ones. A similar trend has been identified in European and American astronomical letters journals.[2] Despite editors' avowed intent to favor short communications over long ones, the average letter is 30 percent longer today than it was ten years ago. The obvious questions are why, and can anything be done about it? Determining just how widespread the phenomenon is and how it correlates with other properties of the papers may suggest answers. To quantify the data, I use a words-in-mean-paper-index (WIMPI), whose charm is that it readily permits comparisons among disciplines, decades, and nations. Investigations of the subject can easily be performed by anyone with access to a large library and a strong index finger (at least the latter should probably be your own).

A preliminary investigation of one English-language journal each of physics, mathematics, and chemistry from the United Kingdom, the United States, and Japan (see the figure caption) shows that the trend spotted by Abt and Harris is very widespread. For every journal examined, WIMPI was larger in 1980–83 than in 1950. Increases ranged from 13 to 115 percent, with an unweighted average of 64 percent. This drops to 43 percent, however, if we set the Japanese baseline year at 1955, by which time the effects of the immediate postwar woodpulp shortage had largely disappeared.

The 13% minimum value belongs to *Monthly Notices of the Royal Astronomical Society* and the 115% maximum to *Journal of the Mathemati-*

cal Society of Japan. After renormalizing the Japanese data, *Journal of the Chemical Society London, Journal of the American Chemical Society,* and the American astronomical journals tie at 82 to 85 percent for the largest increases in WIMPI since World War II. Disciplinary and national averages for the percent increases in words per paper during 1950–80 are physics, 27; mathematics, 77; astronomy, 62; chemistry, 93; United Kingdom, 45; United States, 65; Japan, 85 (but only 20 percent since 1955). Every reader can surely find something in these numbers to support his most cherished preconceptions.

Among the letter sections and journals (most of which are not more than 20 years old in their present formats), 10-year increases in WIMPI range from 5 percent (*Journal of the American Chemical Society*) to 38 percent (*Bulletin of the Chemical Society of Japan*) and 20-year increases from 40 percent (*Bulletin of the Chemical Society of Japan*) to 170 percent (*Journal of Physics*). The average values are 26 percent over the past 10 years and 76 percent over the past 20. Some of the changes closely parallel changes in editorial policy; for instance, the mean paper length in *Physics Letters B* rose from 3.6 pages in 1973, when the rule was "maximum length not to exceed three printed pages," to 4.6 pages in 1983 when it read "should normally not exceed."

The net result is a 1960–80 average increase of 76 percent in WIMPI for the letters publications compared to 29 percent for their associated main journals. Such a trend persisting over half a century would give us 12,000-word "letters," equivalent to about 12 pages of *Nature* (whose authors are currently expected to review even rather broad topics in no more than 6). The high prestige associated with getting one's work into many of the letter publications seemingly encourages authors to push constantly at the length limits so as to squeeze in contributions that should really be full papers.

The complete range of data shows several other trends, strongly suggestive of causality. United Kingdom paper lengths shrank (though not as much as total numbers of papers published) during the First World War and the Depression and rose slightly during World War II (while paper numbers, of course, dropped enormously). American paper lengths changed little during these periods. The reader will immediately think of a variety of explanations in terms of the kinds of people available to do scientific research and the resources available to them. For *Physical Review* and *Proceedings of the Physical Society* (= *Journal of Physics*) the 1980 WIMPIs have only just climbed back to the prewar peaks reached in 1920 and 1930 respectively. *Annals of Mathematics* is unique in show-

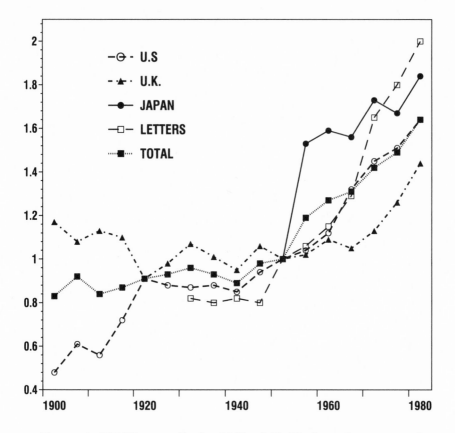

*Changes in WIMPI, normalized to 1950 ≡ 1.00. The journals represented
are: for the United Kingdom,* J. Chem. Soc. *(1900–82),* Proc. Lond.
Math. Soc. *(1903–81),* Proc. Phys. Soc. Lond. *(became* J. Phys.*;
1921–83), and* Mon. Not. R. Astron. Soc. *(1940–82); for the United
States,* Ann. Math. *(1900–80),* Phys. Rev. *(1913–83),* J. Am. Chem. Soc.
*(1920–80), and the American astronomical journals studied by Abt
(1910–1980); for Japan,* Bull. Chem. Soc. Jpn., J. Math. Soc. Jpn., Prog.
Theor. Phys., *and* Publ. Astron. Soc. Jpn., *(all 1950–80). The letters average includes the corresponding short-communications sections and journals plus* Phys. Lett. B, *published in Europe. Data for the earlier years
are generally averages (for example, 1910 ≡ 1908–12), and each point
represents at least 300 papers.*

ing a continuous, nearly linear rise in WIMPI ever since the turn of the
century. Its 1980 papers are, on average, 2.2 times the length of its 1900

ones, reflecting an increasing determination on the part of the editors to accept only papers that present a major new idea in full detail (according to a colleague in mathematics[3]).

The main trends—universal postwar WIMPI creep most heavily concentrated in letter publications, correlations with world economic and political events, responses to editorial policy, and differences among disciplines and nations in (perhaps) expected directions—are all strongly suggestive of an effect largely driven by forces external to the kind of scientific research being reported. This, in turn, provides some hope of moderating the current exponential literature growth. Unfortunately, the sociology of publishing is such that each author's doing what he perceives as best for himself is unlikely to lead to the best possible result for the scientific community as a whole. Perhaps the time has come for some sort of Social Contract among authors, editors, referees, readers, and sponsoring agencies.

Progress Is Not Our Most Important Product

All of us who are old enough to remember are quite sure that many things are not as good today as they were 30 years ago—tomatoes, the post office, operatic tenors, . . . The chief exceptions, of course, are the products of industrial research and development, which are all much better. Or are they? Mitchell Struble and Christ Ftaclas of the University of Pennsylvania report[1] that recent prints of the Palomar Observatory Sky Survey (POSS) contain less information than the older prints and that the chief cause is graininess and specular reflection in the new photographic paper.

The POSS consists of 1870 plates taken on the 48-inch Schmidt telescope from 1949 to 1958, and covers the sky north of $-30°$ declination to a limiting stellar magnitude of about 21 (at the blue end of the spectrum) and 20 (at the red end). A similar survey of the southern sky, being carried out by the UK Schmidt Telescope Unit in Australia and the European Southern Observatory, and a second epoch survey of the northern sky from Palomar are nearly complete. The POSS is a major astronomical tool, used in the identification of radio and x-ray sources; selection of candidate objects for detailed studies; investigations of the morphology of nebulae, galaxies, and clusters; and many other purposes.

The first paper print copies of the POSS were issued in 1954 on Eastman Kodak unicontrast double-weight paper. There have been several subsequent reprintings from the original glass plates, the most recent few on Polycontrast Rapid II RC paper, initially called Ektabrome SC and recognizable by multiple impressions of the words "This paper manufactured by Kodak" on the reverse side (it also feels rather like plastic).

Struble and Ftaclas have studied images of galaxies of varying angular size on both old and new prints (and, for comparison, on one of the much rarer sets of glass copies of the survey). They find, on average, that images on the new prints are smaller (the outer regions of the galaxies have been lost) and rounder (that is, there has been some smearing or loss of angular resolution equivalent to poor seeing) than those on the old prints. The effects are most conspicuous for the smallest images and are sufficient to distort properties of galaxies often measured in morphological studies using the POSS.

I happen to have old and new blue prints of the field including the Crab Nebula and would describe the new print (under × 10 magnification) as looking softer than the old one. For instance, fewer of the discrete filaments near the edge of the Nebula can be resolved from the continuous emission. This would be a blessing in passport photographs for those of us who remember the tomatoes of 30 years ago. Its effects on assorted applications of the POSS deserve further study.

Death Comes as the End

Most of us believe that the frequency with which a scientific paper is cited depends primarily on how useful it is to other workers in the field. Most of us also suspect, however (especially when our own papers go uncited), that there are other factors, related to where, when, and by whom the paper is published. One of these factors is direct personal influence on mentors, colleagues, and students with whom the author communicates. Rao and Vahia[1] have gone so far as to suggest that this personal-influence factor may be sufficiently important to account for the rough proportionality that exists between the number of authors of an astronomical paper and the number of times it is cited.[2] Along the same lines, the fact that astronomers typically write their most-cited papers between the ages of 50 and 60[3] has been attributed by Woltjer[4] to the peaking of personal influence (in the form of directorships, editorships, chairmanships, etc.) during that decade.

The intent here is to probe the effect of personal influence on citation rates. The method is a comparison of the citation histories of papers written by 34 astronomers who died between 1969 and 1982 to the citation histories of papers written by a control group still active in astronomy at the end of 1984. The control individuals were matched as nearly as possible to the index ones in subdiscipline, gender, country of employment, and (most important) year in which career began. This last is important because average citation rates to astronomical papers peak five years after publication and decline monotonically and roughly linearly thereafter, with a half-life near 20 years.[5] The death dates for the sample were constrained at one end by the appearance of the first cumulative issue of *Science Citation Index* including astronomical papers (1965–69) and, at the other end, by the time needed for the astronomical community to be aware of the death when writing papers published in 1984.

The usable sample is not very large, leading to a preliminary conclusion that investigations like this require a larger population (or an older one!) than the world's 7000 research astronomers (median age about 40) to achieve great statistical significance.

The Data

Nearly 200 members of the International Astronomical Union (whose *Bulletin* contains a semiannual necrology) died between 1969 and 1982. Many of them, however, had retired from scientific activities long before; others had devoted most of their later attention to science education, administration, or policy matters, and so had ceased to publish research-oriented papers; and a few, though active to the last, had begun their careers so early in the 20th century that no living individual could serve as a suitable control. As a result, the final sample consists of only 34 astronomers who died with their observing boots well polished and who were still being cited with some regularity in 1984. All but 2 members of the sample appeared as sole or senior author of 2 or more cited papers published less than 3 years before death (and, in some cases, several years after death, owing to the long lead times involved).

For each member of the sample, a control individual was chosen from among the membership of the IAU at the time the index individual died. The control sample is necessary to establish the rate at which citations to papers should drop off with time when the author remains normally influential. Thus first priority was given to matching the year of first published cited paper for index and control astronomers. These agree to within ± 1 year in all cases. Next priority went to matching subdisciplines (theory vs. observation; solar, planetary, stellar, and extragalactic studies, etc.). There were 26 good matches, 4 fair ones, and 4 poor. Gender matched in all but one case. Finally, an effort was made to choose control individuals whose careers took place largely in the same place as the sample individuals' (USSR; United Kingdom; continental Europe; United States/Canada; developing countries; early work in Europe followed by migration to the United States; etc.). Here there were 22 good matches, 8 fair ones (e.g., United States vs. Europe/emigrated to the United States), and 4 poor ones. In only three cases was the match "fair" or "poor" in more than one parameter.

For each index individual, the data consist of the numbers of citations during the periods 1965–69, 1970–74, 1975–79, and 1984 to papers of which he was sole or senior author, excluding obvious self-citations but including all identifiable variants of the name. There were 23,905 citations, an average of 703 per person (high 5279, low 68) and 43.9 per person per year. This is very close to the 1982 average citation rate of 41.4 per year for randomly chosen members of the American Astronomical Society.[6]

For each control individual, only those citations to papers written before the corresponding sample member ended his active career were counted, for the same period and again excluding self-citations. There were 31,891 citations, an average of 938 per person (high 2325, low 217) and 58.6 per person per year. This is very close to the 1982 average rate of 54.4 per year for officers of the AAS.[7]

It is worth noting that the control astronomers have a 33-percent higher average citation rate than the sample ones. This is an artifact of how the groups were chosen. The index astronomers automatically identified themselves by dying. But members of the control group had to be conspicuous enough to be readily identifiable in IAU membership lists as working in the same subdiscipline, etc. as the corresponding index people. The effects of this difference should be removed by the statistical method described in the next section.

The table on page 247 presents only the raw data, which were fairly difficult to extract but fairly easy to analyze. It can be used to check the conclusions that follow or to test other hypotheses.

Results and Discussion

The absolute citation rates range from 1–2 per person per year up to 360 and so are not readily interpretable except as measures of the enormous variability of scientists and science. What we want is a measure of how citation rates for the deceased astronomers have changed in comparison to those for the living ones. The relevant parameter is a double ratio: number of citations in 1984 to papers by deceased astronomer divided by number in last quinquennium he lived through (e.g., 1970–74), divided in turn by the same 1984 to 1970–74 (e.g.) ratio for papers published by the control living astronomer during the active career of the deceased one. Independent of absolute numbers of citations, this ratio should be unity if death does not affect the influence of one's previous research, greater than unity if there is some sort of sympathy factor, and less than unity if the removal of the effect of personal contact diminishes scientific influence.

Let us call this double ratio R. It ranges from 0.03 to 4.88 with a median value of 0.86 for the 34 astronomers considered. The striking result

TABLE 1. Numbers of Citations to Papers by Index (Deceased) and Control (Living) Astronomers 1965–79 and 1984

DATE OF DEATH	INDEX/CONTROL ASTRONOMERS	FIRST/LAST PAPER	NUMBERS OF CITATIONS			
			1965–69	1970–74	1975–79	1984
Oct 1968	Wrubel	1948	57	38	19	1
	King	*1966*	*105*	*111*	*300*	*66*
Nov 1969	Deutsch	1945	256	281	209	23
	Baum	*1971*	*158*	*162*	*118*	*23*
Feb 1970	Henyey	1937	157	138	117	12
	Schwarzschild	*1972*	*506*	*580*	*383*	*54*
Jul 1970	Bernas	1953	109	69	34	2
	Blamont	*1970*	*217*	*140*	*88*	*9*
Dec 1972	Cameron (R. C.)	1961	71	68	45	5
	Cowley (C. R.)	*1972*	*55*	*102*	*55*	*5*
Feb 1973	Bowen	1924	141	146	163	32
	Unsold*	*1973*	*514*	*458*	*292*	*25*
Dec 1973	Hindmarsh	1954	38	141	142	6
	Pagel	*1974*	*176*	*230*	*176*	*5*
Jan 1974	Ferraro	1930	201	123	124	18
	Biermann (L.)*	*1972*	*400*	*298*	*217*	*27*
Feb 1974	Zwicky	1923	415	776	712	105
	Luyten	*1974*	*224*	*172*	*263*	*53*
May 1975	Kiepenheuer	1934	120	104	78	3
	Goldberg (L.)*	*1975*	*523*	*416*	*189*	*16*
Sep 1975	Kukarkin	1934	147	226	417	74
	Ambartsumyan	*1976*	*362*	*298*	*234*	*27*
Nov 1975	Pikelner	1947	240	347	224	19
	Shklovskii*	*1976*	*779*	*750*	*438*	*52*
Jan 1976	Minkowski (R.)	1926	346	405	334	55
	Wilson (O. C.)	*1975*	*314*	*421*	*354*	*75*
Dec 1976	Menzel	1923	238	189	150	13
	Oort	*1976*	*489*	*661*	*576*	*104*
Apr 1977	Limber	1953	99	159	111	12
	Kraft	*1976*	*417*	*501*	*493*	*84*
Sep 1977	Huang (S. S.)	1937	157	215	248	32
	Kopal	*1977*	*953*	*561*	*711*	*100*
Nov 1977	Chalonge	1934	61	55	49	8
	Whitford	*1976*	*102*	*147*	*203*	*25*
Feb 1978	Thackeray	1933	168	244	258	51
	Stromgren (B.)*	*1978*	*356*	*349*	*224*	*28*
Mar 1978	Lallemand	1935	83	73	39	7
	Kron (G. E.)	*1976*	*176*	*224*	*310*	*30*

*Since deceased.

continued on page 248

TABLE 1 *(continued)*

DATE OF DEATH	INDEX/*CONTROL* ASTRONOMERS	FIRST/LAST PAPER	NUMBERS OF CITATIONS			
			1965–69	1970–74	1975–79	1984
Jun 1978	Kaplan (S. A.)	1945	190	337	352	41
	Sobolev	*1979*	*408*	*602*	*491*	*84*
Apr 1979	McCuskey	1938	64	80	95	11
	Popper	*1976*	*139*	*140*	*246*	*24*
Sep 1979	Syrovatskii	1953	111	215	316	43
	Ivanov (V. V.)	*1980*	*110*	*248*	*329*	*28*
Dec 1979	Payne-Gaposchkin	1925	103	134	205	38
	McCrea	*1977*	*188*	*201*	*274*	*35*
Mar 1980	Myerscough	1962	27	23	17	1
	Jordan (C.)	*1972*	*62*	*278*	*402*	*46*
Apr 1980	Bullard	1930	367	641	528	75
	*Cowling**	*1978*	*417*	*297*	*270*	*42*
Apr 1980	Johnson (H. L.)	1947	1422	1777	1794	286
	Code	*1980*	*125*	*271*	*406*	*51*
Dec 1980	Wyatt	1950	30	23	41	10
	McNamara	*1977*	*121*	*97*	*100*	*5*
Mar 1981	Mueller (R. F)	1960	144	178	152	22
	Anders	*1978*	*467*	*544*	*804*	*51*
May 1981	Tinsley	1967	11	86	464	69
	Trimble	*1981*	*28*	*187*	*311*	*66*
Oct 1981	Serkowski	1956	99	235	382	48
	Low	*1980*	*343*	*571*	*337*	*29*
Dec 1981	Whelan	1970	0	31	89	20
	Pringle	*1979*	*0*	*119*	*449*	*57*
Feb 1982	Neyman (J.)	1923	374	331	471	56
	*Opik**	*1980*	*461*	*503*	*382*	*38*
Aug 1982	Bappu	1951	30	48	38	18
	Blanco (V. M.)	*1981*	*121*	*236*	*363*	*56*
Nov 1982	Linfoot	1943	141	100	104	25
	Gascoigne	*1972*	*116*	*133*	*126*	*17*

*Since deceased.

emerges when we consider R for astronomers who ceased work in different years. Among the 8 who stopped publishing before 1974, R's < 1 outnumbers R's > 1 by 7 to 1. For the group who died between 1974 and 1978, R's of less than one again lead by 11 to 4, but among the 11 astronomers who ceased work most recently (1979–82), there are 9 R's > 1 and only 2 smaller ones. The medians for the three groups are 0.65, 0.87, and 1.45. One's impression is of a brief outburst of sympathy, followed by gradual forgetting of the deceased's contributions, at least in comparison with those of similar but still active astronomers. A "sympathy period" can also be identified for the first two groups, who died between 1969 and 1978. An analysis like the present one carried out in the quinquennia immediately following their deaths would have found R's greater than unity leading by 5 to 3 in the first group and 9 to 6 in the second.

A number of colleagues with whom these results have been discussed have claimed that they are exactly what one expects in the wake of death—a brief period of memory and sympathy followed by gradual forgetfulness. They are probably right, and perhaps the main lesson is that there is a lot of human nature in all of us.

The amounts by which the median R's for the various groups deviate from 1.0 suggest that about 40 percent of astronomical citations may be mediated by some degree of personal influence. Rao and Vahia[8] reach a rather similar conclusion, using a very different data base and method.

Young Versus Established American Astronomers

The younger generation has been either going to the dogs or repairing the damage done by its elders (depending on which side of the fence you sit on) at least since the time of Quintus Horatius Flaccus (*c*. 25 B.C.E.).[1] One hears, for instance, that young astronomers are more prompt in publishing their data, but less likely to read the archival literature, than their predecessors. Neither of these is easily susceptible of quantitative testing. Some other possible differences are testable, given data on enough members of each generation. Chance simultaneous service on several committees placed on my desk résumés of 269 mostly young (median Ph.D. 1982) astronomers who have recently applied either for tenure-track positions in astronomy/astrophysics in the United States or for election to membership in the International Astronomical Union from the United States.

Many questions could be asked about these young astronomers. The two addressed here arose in discussion with colleagues: (a) Is the United States becoming increasingly dependent on immigrant scientists? and (b) Is the average length of time taken to earn a Ph.D. increasing? In comparison with a sample of 304 established astronomers, the answers turn out to be no and yes, respectively.

The Samples of Applicant and Established Astronomers

Applications for IAU membership from the United States were filed by 184 people in 1988. (The opportunity arises every three years.) Two recently advertised assistant professor positions in astronomy or astrophysics attracted 92 applications, 7 from people who had also applied for IAU membership (and who were analyzed as part of the IAU sample).

Thus, the "applicant astronomer" group contains 269 people. The median applicant received a Ph.D. in 1982, with 68% between 1978 and 1986. A few of them did not provide information on places of birth or on dates and places of degrees or never received formal Ph.D. (or B.A./B.S.) degrees. Thus, some total numbers in the tables and figures are slightly smaller than 269.

The two applicant samples are statistically not quite the same. The ground rules for IAU election from the United States include a record of publication for at least three years past the Ph.D. or equivalent and citizenship or permanent residence. In effect, the academic community and the Immigration and Naturalization Service ask for rather similar characteristics in assistant professors, but the requirements are not formalized and so the job-applicant pool includes a few very recent Ph.D.s and a few who have never worked or studied in the United States The differences are, however, small, and the two groups are largely treated together in the next section.

A comparison group of "established astronomers" had to be created. This was done by selecting (a) all current officers, councilors, and committee members of the American Astronomical Society from its 1988 directory, (b) all living prize winners and former officers and councilors of the AAS, and (c) a subsample of people elected to the IAU at or before its 1982 General Assembly who listed U. S. addresses at the time of their election, taken from every 10th page of the 1983 IAU directory. About 10 astronomers who had not spent a significant part of their careers in the United States were eliminated from subsamples (a) and (b). This yielded 418 people reasonably describable as established American astronomers. Data on their places of birth and dates and institutions of academic degrees were then extracted from the 16th edition of *American Men and Women of Science (AMWS)*.There were no listings for about one-quarter of them, reducing the usable sample to 304. The median established astronomer received his Ph.D. in 1962.5 (68 percent between 1952 and 1972). Incomplete data in some AMWS listings slightly reduce some of the tabulated total numbers.

One worries about systematic differences between the listed and unlisted astronomers as a source of bias. The unlisted are, on average, younger than the listed (both because of the way *AMWS* selects its biographees and because such volumes have, in recent years, proliferated to the point where no one can fill out all the forms that arrive on his desk). This merely separates our applicant and established samples a bit more in time and is all to the good. The unlisted are not, on average, employed at more or less prestigious places than the listed and show no other evidence of higher or lower competence or repute that might somehow be correlated with earning degrees rapidly or slowly. Finally, foreign-born astronomers may be slightly overrepresented among the unlisted (either because they find a form labeled "American something or other" unattractive or because they are just not questionnaire minded). The statistical effect of this is addressed briefly in the next section.

Characteristics of the Samples

Tables 1 and 2 indicate nations of birth and first degree (B.A./B.S.) of the applicant and established astronomers. Ph.D.s were earned in either the United States or the country of first degree, except where indicated. Present employment or residence is in the United States except where indicated. Country names are those in effect at the time of first degree or birth.

A large majority of both groups is U.S. born and educated. But 27% of the applicants (72 of 269) and 23% of the established astronomers (71 of 304) are foreign born. In addition, 21% of the applicants (57 of 269) and 19% of the established sample (58 of 304) received their first degrees (B.A./B.S.) abroad.

The applicant and established groups are really even less different in this parameter than they at first appear. Four or five of the job applicants have never actually worked or studied in the United States and may well never, in fact, do so, reducing the real foreign-born fraction here to about 25%. Among established astronomers, on the other hand, a disproportionate fraction of the foreign born were lost through not being listed in *AMWS*. Most of the 114 unlisted people are at least personal acquaintances, of whom 25 were definitely and 8 probably born outside the United States. Thus the total group of established astronomers is also about 25% (104 of 418) foreign born.

The corresponding numbers may well be higher or lower or changing for other sciences, but astronomy in the United States has been remarkably stable in drawing about one-quarter of its practitioners from other countries. The distribution among countries of origin also looks about as stable as shifting politics and statistics of small numbers would permit. In particular, the United Kingdom has contributed about 5 percent of both the applicant (13 of 269) and established (15 of 304 or 24 of 418) astronomers. It is perhaps significant that the colleague who originally suggested that American astronomy was becoming rather parasitic is British.

TABLE 1. Countries of Origin of Applicant Astronomers

BIRTH	B.A./B.S.	NUMBER	
USA	USA	197	(PhDs 1 UK, 1 GFR; 2 employed Netherlands, 1 UK, 1 Israel)
UK	UK	13	
Canada	Canada	10	(1 Netherlands PhD, 1 employed Canada)
	USA	1	
Israel	Israel	3	(1 GFR PhD)
	USA	1	
Taiwan	Taiwan	2	
	USA	2	
Argentina	Argentina	3	
India	India	3	(1 employed Italy)
Mexico	Mexico	3	(1 UK PhD; 1 employed Canada)
Poland	Poland	3	(1 refugee)
Germany	DDR	1	
GFR	GFR	1	
	Spain	1	
Australia	Australia	2	(1 UK PhD)
Japan	Japan	2	(1 employed Japan)
Finland	Finland	1	
	USA	1	
Iran	USA	1	
Belgium	Belgium	1	(employed Switzerland)
Chile	Chile	1	(employed Switzerland)
Italy	Australia	1	
Lebanon	Lebanon (Am Univ)	1	
Netherlands	Netherlands	1	
New Zealand	New Zealand	1	
PR China	PR China	1	
Romania	Romania	1	
Yugoslavia	Yugoslavia	1	
Czechoslovakia	USA	1	
Hong Kong	USA	1	
Hungary	USA	1	
Sweden	USA	1	
Switzerland	USA	1	
Thailand	USA	1	
Venezuela	USA	1	
Stateless	USA	1	

TABLE 2. Countries of Origin of Established Astronomers

BIRTH	B.A./B.S.	NUMBER	
USA	USA	232	(PhDs 2 UK, 2 Australia, 1 Canada, 1 Israel; 2 now in Canada)
	Canada	1	
UK	UK	13	
	Australia	2	
	USA	1	
Canada	Canada	8	
	USA	1	(now in Canada)
Germany	Germany	1	
	GFR	3	
	Israel	1	
	USA	3	
Netherlands	Netherlands	6	(1 now in GFR)
Australia	Australia	1	
	UK	1	
	USA	1	
India	India	3	
Italy	Italy	2	
Ireland	Ireland	1	
	UK	1	
Israel	Israel	1	
	USA	1	
Switzerland	Switzerland	1	
	USA	1	
Hungary	Hungary	1	
	USA	1	
China	USA	2	
Argentina	Argentina	1	(PhD Italy)
Denmark	Denmark	1	
Mexico	Mexico	1	(now in GFR)
Greece	Greece	1	
South Africa	South Africa	1	
Czechoslovakia	Canada	1	
Hong Kong	Canada	1	
Morocco	Canada	1	
Austria	Australia	1	(PhD UK)
Danzig	USA	1	
France	USA	1	
Latvia	USA	1	
Monaco	USA	1	
Russia	USA	1	

The length of time required to earn a Ph.D., in contrast, differs significantly between the two samples. Figures 1 and 2 show the distributions of numbers of years between B.A./B.S. and Ph.D. for the applicant and established astronomers, respectively. The shapes are rather similar, but the two histograms are displaced by a year, the applicant plot peaking at 6 years and the established one at 5 years. In both samples, intervals of 10 or more years typically reflect the intervention of some major nonacademic activity. Military service is the commonest in both groups, but the range includes child raising, the ministry, and beachcombing.

What are we to make of the difference? The first, happy, reaction is that the established community has been through a filter that has removed those who were slow to earn PhD's because they were less interested or less productive than average. This would be a comforting thought, especially to members of committees who decide the number of years a student is permitted to spend on graduate work. Unfortunately, the breakdown of the data shown in Table 3 suggests that this first thought is not the whole story. Instead, the dominant effect seems to be a gradual increase with time in years required to earn a Ph.D. in both samples, separately and in their sum.

Table 3 lists numbers of applicant, established, and total astronomers who took from 2 to ≥ 11 years from B.A./B.S. to Ph.D. The median in each column is in underlined type and can be seen to have increased from 4 years for PhD's earned before 1954, to 5 years for those between 1954 and 1974, and to 6 years after 1974. Another way to look at the numbers is as the fraction of Ph.D. recipients who took 7 or more years. This increases monotonically from 0.18 (1931–53) through 0.27 (1954–74) and 0.36 (1975–79) to 0.48 (1980–88). The change has not been an increase in very long (≥ 10 year) times, but rather a gradual replacement of 4- and 5-year degrees by 6- and, increasingly, 7- and 8-year degrees. The change would look even more extreme without the U.K. immigrants, who contributed disproportionately to the shorter times in the applicant (but not in the established) sample. The 4-year Ph.D. may still be the norm in many of our minds, but it has not been the average for more than 30 years.

About three-quarters of the astronomers in both samples received M.A. or M.S. degrees between the B.A./B.S. and Ph.D., very typically two years after the first degree. It is not clear, however, that the intervals between B.A./B.S. and M.A./M.S. or between M.A./M.S. and Ph.D. contain any additional information, since many institutions award these de-

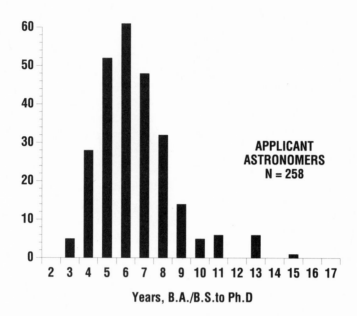

FIGURE 1.   *Distribution of numbers of years between B.A./B.S. and Ph.D. for applicant astronomers.*

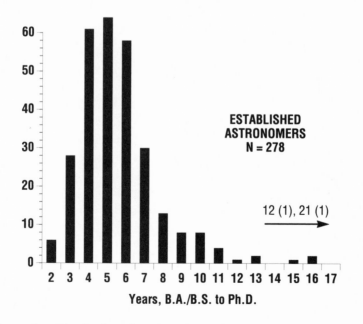

FIGURE 2.   *Distribution of number of years between B.A./B.S. and Ph.D. for established astronomers.*

TABLE 3.   Distribution of Years Between B.A./B.S. and Ph.D.
for Applicant (A), Established (E), and Total (T) Astronomer Samples Subdivided by Year of Ph.D.
(median of each subsample is underlined)

YEARS B.A./B.S. TO PH.D.	PhD Recieved In																				
	1931–53			1954–62			1963–68			1969–74			1975–79			1980–83			1984–88		
	A	E	T	A	E	T	A	E	T	A	E	T	A	E	T	A	E	T	A	E	T
2	0	4	4	0	1	1	0	1	1	0	1	1	0	0	0	0	0	0	0	0	0
3	1	13	14	0	9	9	0	3	3	1	2	3	0	0	0	1	0	1	2	0	2
4	2	_19_	_21_	2	9	11	0	14	14	4	11	15	5	7	12	8	0	8	5	0	5
5	1	8	9	0	_16_	_16_	2	_20_	22	8	_17_	_25_	8	_5_	13	23	0	23	12	0	12
6	0	11	11	0	12	12	_2_	12	14	3	11	14	_13_	3	_16_	_13_	0	_13_	_28_	0	_28_
7	0	2	2	1	7	8	0	7	7	2	11	13	6	3	9	16	0	16	23	0	23
8	0	1	1	0	5	5	1	5	6	2	0	2	4	1	5	13	0	13	12	0	12
9	0	3	3	0	1	1	0	1	1	2	1	3	3	1	4	4	1	5	5	0	5
10	0	2	2	0	1	1	1	3	4	1	0	1	2	1	3	0	1	1	2	0	2
≥11	0	5	5	0	4	4	0	2	2	1	2	3	3	0	3	7	0	7	2	0	2
TOTAL	4	68	72	3	65	68	6	68	74	24	56	80	44	23	67	85	2	87	91	0	91

grees more or less automatically to graduate students who have completed a certain number of units, passed (or failed!) their qualifying exams, or otherwise simply made normal progress.

One final datum is readily available. Of the applicant astronomers, about 9 percent are women (18 of 184 for the IAU and 7 of 85 for the assistant professorships) as are 11 percent of the established astronomers (30 of the 304 with *AMWS* listings and 17 of the 114 not listed). This 10% level seems to be typical of many current populations. A count of every 6th page (28, 34, 40, . . . 94) of the 1988 AAS directory found 88 to 92 women among 817 members, or 11 percent. The uncertainty reflects 4 probably but not uniquely feminine given names among members not personally known to me. Among other possible samples, authors in *Annual Reviews of Astronomy and Astrophysics* under the present editor are 9 percent (29 of 315) female, and 1987 *Publications of the Astronomical Society of the Pacific* authors are at least 10 percent (40 of 415) female, representing a lower limit owing to the assumption that all authors are men who give only initials and are not known to be women.

Other subsets over- or under-represent women, who make up 18 percent of AAS officers and committee members in both the 1988 and 1984[2] directories but only 5 to 6 percent of astronomers at prestigious institutions[3] and of living winners of AAS prizes excluding the Cannon Prize (8 of 132 and 6 of 117, respectively).

What about secular trends? Counting every 6th page of the 1973 AAS directory yielded 33 women among 530 members (6 percent). *Annual Reviews* authors under the previous editor (1962–73) were only 4 percent female (9 of 226), and the 299 PASP authors in 1972 included 19 (6 percent) women. The difference between about 10 percent now and about 5 percent then is probably real, but it is not part of a monotonic, long-term trend: a photograph of the AAS founding members in Williams Bay in 1899 included 9 women in a group of about 92.

Potential Sources of Error

The 304 established astronomers were selected from American Astronomical Society officers and prize winners and International Astronomical Union members and so are largely people who really do astronomy (research, advanced teaching, organizing) for a living. The sample is thus more or less likely by definition representative of established astronomers and includes 20 to 30 percent of full-time American ones with pre-1975 PhD's.

The chief remaining possible source of error here is the foreign-born component of established astronomers not listed in *AMWS*. One does not often go up to a colleague and say something like "You have a beautiful midwestern accent and seem to know nothing whatever about the rest of the world. Where were you born?" Still less often is the answer "Vladivostok." But it happens: one (though only one) of the 304 listed astronomers who I had always assumed was native born, despite a somewhat exotic name, is not. Thus the immigrant fraction may be a little larger than 25 percent.

The 269 young astronomers were self-selected through applying for IAU membership or recently advertised tenure-track jobs. They have, therefore, declared their intention to remain or become part of the professional American astronomical community. Given that something like 40 to 50 new PhD's seek first positions in astronomy each year, the sample must include one third to one half of the 1978–86 cohort and a somewhat larger fraction of those who will, in fact, remain part of the community, since there has already been some filtering.

It is possible that those who have chosen not to join the IAU at this time (and have applied for other jobs or are content with their present ones) are somehow different from the present samples in ways that would undermine the main conclusions. They might, for instance, have earned their PhD's faster on average or include more immigrants. A review of acquaintances not in the sample does not support this possibility. In any case, nonapplicants are quite unlikely to be more typical representatives of future astronomers than the applicants.

Conclusions

Data on samples of young and established American astronomers (median PhD's 1982 and 1962.5, respectively) have been examined for countries of origin and length of time elapsed between first degrees (B.A./B.S.) and Ph.D. Both groups are about 25 percent foreign born, with the United Kingdom contributing one-fifth of the immigrants, or 5 percent of both samples. Both samples are about 10 percent female, as is the present general AAS membership. Times elapsed between B.A./B.S. and Ph.D., on the other hand, are not constant. The median has increased from 4 years before 1954 to 5 years for 1954–74, to 6 years for degrees earned after 1974. Of the youngest astronomers (PhD's 1980–88), 48 percent took 7 or more years to earn their PhD's. Though we may think of 4 years in graduate school as a desirable standard, it has not been average for more than 30 years, and 7- and 8-year degrees are by no means uncommon today.

Prestigious Start

If you really want a career in scientific research, then get your Ph.D. from the most prestigious university that will take you. This conclusion (which should surprise very few) follows from data collected by the Astronomy and Astrophysics Survey Committee for a report to the U.S. National Academy of Sciences on Astronomy and Astrophysics in the 1990s. To be precise, astronomers who received their doctoral degrees from a university that ranks repeatedly among the "top three" American astronomical Ph.D. programs[1] continue to publish in the field about twice as long as those with degrees from a university that ranks in the "second ten." In addition, they are about twice as likely to have current jobs that primarily involve research or advanced teaching at observatories and/or PhD-granting universities.

One can imagine a variety of reasons for the difference. The more prestigious institution may genuinely provide a better education, or it may skim the cream from the available pool of entering students. Or its name may serve as a lubricant to smoothe the paths to research grants, acceptance of papers, and good jobs for its alumni, at least early in their careers. The data say nothing whatever about the relative importance of these or other causes for the phenomenon, though I am privately of the opinion that the third factor is not negligible.

The sample populations include 106 recipients of Ph.D.s between 1952 and 1988 from the high-ranking institution (hereafter P) and 94 recipients of Ph.D.s between 1966 and 1988 from the lower-ranking one (NP hereafter). Figure 1 shows their publication histories. The horizontal axis is number of years since receipt of doctoral degree. The vertical axis is the fraction of all astronomers who had had their degrees for at least N years who were still publishing papers listed in the semi-annual issues of *Astronomy and Astrophysics Abstracts*[2] at least N years after their degrees. Upper points pertain to P and lower ones to NP, and the differences are large enough to require no words to guide the eye.

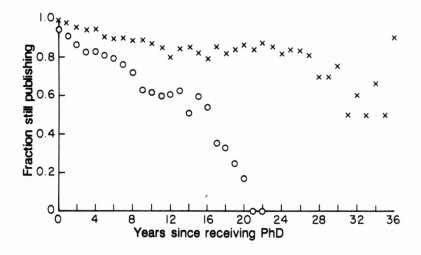

FIGURE 1.   *Fraction of astronomers who have held Ph.D.s for at least N years who are still publishing research papers N years after their degrees. Upper points = people from the prestigious university, P; lower points = people from the nonprestigious university, NP. Neither set of points starts out a unity because a few people from each place never published anything. Numbers drop below 10 people at 18 years for NP and 28 years for P.*

The doctoral program at P is the older one, and it might be hypothesized that the difference shown in Figure 1 is a result of good jobs being easier to get in the past (back when our towers were made of real ivory, rather than plastic, and all university students were above average). As a test of this hypothesis, Figure 2 shows the publication histories of post-1966 degree recipients from the two universities (85 from P, 94 from NP). Indeed the difference between the two is smaller than in Figure 1, but the P astronomers are still about twice as likely to continue to publish for more than 18 years (when both samples drop below 10 members) than are the NP ones.

Current employment patterns of the two groups are also significantly different. There are members of each group in each of nine categories: (1) astronomy research and teaching at observatories and PhD-granting universities; (2) astronomy research at government and industrial laboratories (such as Los Alamos National Laboratory and AT&T Bell Laboratories);

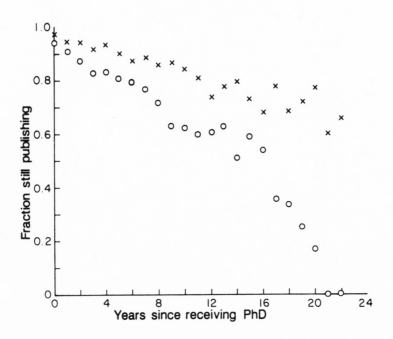

FIGURE 2. *Fraction of post-1966 degree recipients who are still publishing in astronomy as a function of number of years since the Ph.D. Upper points = P alumni; lower points = NP alumni. Numbers drop below 10 people between 18 and 19 years post-Ph.D. for both.*

(3) research in other sciences; (4) scientific support services (data processing, hardware and software development); (5) industrial employment; (6) government employment (generally science administration); (7) teaching at places with no Ph.D. program in astronomy; (8) employment outside science; and (9) lost to follow-up and not publishing in astronomy (2 from P, 3 from NP). The most striking difference is that 65 percent of the P astronomers, but only 32 percent of the NP ones, have jobs in category 1. A large majority of incoming graduate students at both places describe their career goals as research and advanced teaching in the field. Most graduate programs concur that their primary purpose is the production of the next generation of researchers. Thus the P astronomers can be described as more successful than the NP ones by their own standards.

Life being somehow easier in the past is not the primary cause. An artificial sample of P astronomers, age-matched to the NP ones, has 66 percent of its members in the most desirable category of employment, just the same as for the total P sample. Curiously, this does mean that things are getting worse. People continue to diffuse out of the astronomical research community over their entire careers. But they almost never come back (though there is diffusion into the field of Ph.D.s from physics and other sciences). No one in either sample resumed publishing after a hiatus of as long as 5 years. And the few returns to jobs in categories 1 and 2 occurred only when the excursions had been brief ones to positions in science administration. Thus we can expect that, 10 or 20 years from now, the fraction of the 1966–88 P astronomers still in research and advanced teaching will be less than 65 percent and so smaller than the fraction of the total P sample now engaged in these activities.

The moral of the story is thus twofold: it still pays to go to the most prestigious graduate institution you possibly can, but the gold paving, even on the streets of P, is not as thick as it used to be.

Self-Citation Rates
in Astronomical Papers

Examination of the 10,910 references contained in 496 astronomical papers published during January 1983 indicates that 1645 of them ($15.1 \pm 0.5\%$) were self-citations, in the sense that citing and cited papers had at least one author in common. This is rather higher than has generally been adopted in analyses where the self-citation rate is needed as an input.[1] The discrepancy arose largely because the 6.4% self-citation rate most often referred to actually counts only pairs of papers with the same senior or sole author.[2]

The table on page 266 lists the publications examined, numbers of papers and citations, and percentages of self-citations. Where more than one issue of a journal carried a January 1983 date, all were used. Where the January/February/March issue covered two or three months, every other or every third article, beginning with the first, was used. Various regional averages are shown. Since the International Astronomical Union now publishes about 12 volumes of colloquia, symposia, and proceedings each year, the volume received in January 1983 (Symposium No. 99) was also included.

Variations among journals and regions are surprisingly small. The investigation was originally intended to test the hypothesis that self-citations are significantly less common in the *Astrophysical Journal* than in other publications. This seems not to be the case. Another rather constant number is the average number of self-citations per article. This falls between 2.0 and 3.4 for nearly all journals and regions represented by more than a few papers, except for the *Astrophysical Journal Supplement* (at 10.2). Several of the more extreme values derive from quite small numbers of papers and should not be regarded as significant.

It is, obviously, impossible to say what these data mean in isolation, or whether the current rate of self-citations in astronomy is a signature of a

Numbers of Citations and Self-Citations in
Astronomical Papers Published in January 1983

JOURNAL VOL., NO.	NO. OF PAPERS	NO. OF CITATIONS	NO. OF SELF-CITATIONS	PERCENTAGE OF SELF-CITATIONS
Astrofiz 19, 1	6	84	15	17.9
Sov AJ 27, 1	15	229	33	14.4
Sov AJ Lett 9, 1	11	156	37	23.7
Soviet total	**32**	**469**	**85**	**18.1**
AN 304, 1	5	82	8	9.8
* Acta Astr 33, 3–4	2	84	17	20.2
BAC 34, 1	4	74	26	35.1
East. Eur. total	**11**	**240**	**51**	**21.3**
PASJ 35, 1	5	112	17	15.2
J Ap Astr 4, 1	3	51	4	7.8
Chinese Ast	6	57	5	8.8
Proc ASA	2	76	6	7.9
BAS India 11,1	3	44	4	9.1
Asia–Pacific total	**19**	**340**	**36**	**10.6**
AAp 117	54	1511	180	11.9
AAp Sup 51,1	19	325	55	16.9
ASS 89	41	618	89	14.4
** Sol Phys 83 (Feb)	14	265	47	17.7
Moon & Plan 28, 1	6	102	15	14.7
West Eur. total	**134**	**2821**	**386**	**13.7**
Nature 301, 1–4	9	120	23	19.2
MNRAS 202, 1	32	928	120	12.9
Obs no 1053	2	31	2	6.5
Plan Sp Sci 31, 1	12	315	59	18.7
U.K. Total	**55**	**1394**	**204**	**14.6**
Total journals	**251**	**5264**	**762**	**14.5**
IAU Symp 99	**82**	**1157**	**223**	**21.0**
Total ex. N. Am.	**333**	**6421**	**985**	**15.3**

continued on page 267

healthy, growing science or of an overly inbred one. Comparisons with other sciences and other eras might aid interpretation. The astronomical rate, at least, has apparently been quite stable. The first volume of the *Astrophysical Journal* in which references are tabulated at the end of papers rather than in footnotes (vol. 121, 1955) contains 84 papers with 1119 ci-

**Numbers of Citations and Self-Citations in
Astronomical Papers Published in January 1983** (continued)

JOURNAL VOL., NO.	NO. OF PAPERS	NO. OF CITATIONS	NO. OF SELF-CITATIONS	PERCENTAGE OF SELF-CITATIONS
JRAS Can 77, 1	2	39	3	7.7
** Rev Mex A&A 8,1	2	38	6	15.7
Science no. 4080–83	3	19	9	47.4
Meteoritics 18,1	3	41	4	9.8
Ap Lett 23, 2	3	91	8	8.8
Icarus 53, 1	15	363	60	16.5
PASP 95, 1	19	303	56	18.5
AJ 88, 1	18	456	60	13.2
ApJ Sup 51, 1	6	329	61	18.5
ApJ Lett 264	13	269	42	15.8
N. Am. ex. ApJ	**84**	**1948**	**309**	**15.9**
ApJ 264	79	2541	351	13.8
Total N. Am.	**163**	**4489**	**660**	**14.7**
GRAND TOTAL	**496**	**10,910**	**1645**	**15.1**

*January 1983 issue never received. **January 1983 issue contains conference abstracts.

tations and 166 self-citations (14.8 $\pm$ 1.5%). This stability perhaps reflects the compensating influences of increase both in the average number of authors per paper and in the size of the author pool that could, in principle, be cited. Comparable data on other sciences could readily be collected, but not by the present author, who finds picking a few names out of a long list fairly easy when most of them are familiar but quite impossible when they are not.

In the absence of other information, there is no *a priori* reason to conclude that the 15-percent self-citation rate in astronomy is too high (or too low) or indicative of anything in particular about the psychology of authors. Even anomalously low or high rates need have no profound implications. Zero percent is typical and reasonable for a new Ph.D. publishing the results of his thesis, while 30 percent or even more is not unreasonable for a more senior author who has done pioneering work in one or more areas (H. L. Johnson and O. Struve are examples from the 1955 sample). The table reports a measurement process that is itself likely to affect the phenomenon. As studies of citation rates and their use to evaluate journals, institutions, scientists, and telescopes proliferate, authors cannot help but become conscious of the citation process and be tempted to modify their usual practices.

I am indebted to Dr. Helmut Abt for his usual thoughtful review of the initial version of this note.

ANTIC MUSE

Brief Encounter with a Facilities Manual

ABSTRACT: Large observatories which make some time available to the general astronomical community normally provide instruction manuals for the main telescopes and the focal plane instruments or receivers. These are, as a rule, prepared by the person who developed the apparatus and who generally already knows a good deal about it. Unfortunately, this is not always also true of the first-time user. We report a short extract from a typical manual and a novice observer's interaction with it.

Received 1943 November 15; accepted 1968 April 12

Step 12: Backstep Reticulator

Happily the reticulator, unlike the correlator, the dark current counter, and the background calibrators described in Steps 2, 5, and 11 respectively, carries its name upon its face. It is a black near-cube, about a foot and a half (0.5 m) high, attached by cables to three of the possible four nearest neighbor boxes and to five of the possible sixteen next-nearest neighbors.

On its face are twelve dials in two rows, three switches, and two screens, one long and narrow and apparently intended to display alphameric information, and one square and presumably intended for graphical information. Both are currently dark and remain so throughout. Discouragingly, none of the dials is labeled "backstepper." In fact, none of them is labeled at all, exept the left-handed dial in the top row, above which a small gummed paper sticker sternly declares "These are *loga-*

rithmic units." The possible settings are A, B, C, D, E, and K_1. The dial is set at D, which seems to be as good a logarithm as any. And the switch near the rectangular screen is held firmly in the down position by a piece of adhesive tape on which faint blue lettering admonishes even more sternly "Do not attempt to rotate reticulator slit unless you have been checked out by JCM." (Atempt is misspelled.) Novice observer begins to suspect that some form of checking out (preferably out of the observatory dorm) might be a good idea.

With fading hopes and dangling participles, further enlightenment is sought from the primary instruction sheet, where an earlier observer has penciled the remark "Reticulator must be on before backstepping." Novice observer backsteps on tiptoe out of the control room all the way to the parking lot and heads for home, where high priority is given to the task of revising *curriculum vitae* to describe primary research interest as astrophysics rather than astronomy.

Acknowledgments: Several major observatories have made significant contributions to this study at a variety of wavelengths.

Astronomical Conferences

As Seen by a Participant

The world is full of things I don't understand (including galaxy formation, peculiar A stars, and the solar neutrino problem). But high among them ranks the kinematics of astronomers at international meetings.

Frankly, being in a foreign country where you don't speak the language, haven't mastered the currency, and can't read the maps can be a little scary. But somehow, once you see another astronomer in the airport transit lounge, it's all OK. He may not speak the language either. It may even be Professor Hjalmar Sciatti who, to first order, doesn't seem to speak any language at all and whose colloquia you normally avoid like the plague. For the moment, though, you are buddies, and if you ever manage to find the Vigan Bhavan, you will automatically stand in the registration line together. In this sense, the astronomical community is still a small town or extended family. And one of the purposes of the triennial General Assembly of the International Astronomical Union is to welcome new members.

By the third day of a meeting, most of us will have staked out our favorite seats in the main lecture hall and be looking askance at anyone who dares to trespass thereon. (Mine is in the front left corner, and I'll thank you to keep out of it.) Even earlier, we will have marked out our paths from hotel to conference center to food and back so reliably that we repeatedly encounter the same few people no matter how many participants there may be. Before the first formal lecture of the Baltimore General Assembly, I had already run into Hubert Reeves three times, and never did give up hoping that we would eventually be going the same direction and so have a chance to say more than "ciao."

As Seen by a Member of the Organizing Committee

Why is the registration fee so high when all I get is a canvas folder and some coffee? You may have asked yourself that question several times in planning your trip to an IAU General Assembly. You may even have asked a member of the Local Organizing Committee or the session chairman who invited you to speak. I hope our answers were at least polite, though I cannot promise.

Well, what do you get? The real budget and which inputs correspond to which outputs are invariably so complicated that only the business manager understands all the details well enough to keep the rest of us out of jail. Very roughly, though, grants totaling a bit more than half a million dollars from national agencies will cover basic expenses—rental of the convention center facilities, audio-visual equipment, security, many of the necessary coordinating and consulting services (like the company handling the registrations), coffee, and a well-spent subsidy for young astronomers from all over the world, many attending their first major meeting. Hotel rooms, dormitories, and tours for family and friends are budgeted separately and do not enter into this equation.

A conference with only these basics would, however, be very bleak indeed—not just no parties, but also no telephones, no programs or newspapers, no mailboxes, and no signs to tell you where things are. And none of the amenities is free. For instance, the carpeting and draperies that make quiet conversation and walking simultaneously possible in the convention center set you back about $3 (1988 dollars) per day.

In fact, income would fall very short indeed of equaling outgo without generous donations from organizations (industrial, educational, and civic) interested in supporting astronomy. And now, if you will get out your pocket calculator, you can think of your registration fee as having bought you an $18.50 portfolio (including the contents), $17.60 in coffee breaks, $30.25 in carpeting and other amenities, a $40.35 opening reception (this includes admission to both the Aquarium and the Science Center as well as the food and drink and water taxies), a $30.30 symphony concert, a $15 evening of American music, and a $45 closing banquet (after subtracting the $10 separate ticket price). Oh yes. There are also ten 79-cent newspapers, for which these remarks were originally written (so I hope nobody wants his money back).

As Seen by the Local Hosts

Is red, white, and blue too patriotic? Does the "20" have to be in roman numerals? What if it rains and the canvas gets wet? These were the local hosts hard at work a year or so before the General Assembly, settling the burning issues of the appearance of the Union logo and the conference portfolios. No, that isn't quite all such groups do. But, in accordance with traditional committee practice, the amount of time devoted to discussing each issue is roughly proportional to the reciprocal of the cost.

The real planning for one triennial assembly begins even before the previous one ends. I can't add up the total number of telephone hours put in, but while some hosts were collecting competitive bids from bus companies, others were negotiating with hotels, printers, caterers, carpenters, and the convention center. What is really needed, how much is it going to cost, and how could it possibly be made to cost less? Special sympathy should be extended to the brave subcommittee who previewed the menus for the closing banquet and other social events.

The hardest part always is guessing how many astronomers and guests to plan for. Registration fees taken in are, of course, linear in the number of participants. But some of the costs are not, and we estimated that we would break even at 2000 and be able to have pastries with coffee at 3000. The truth fell somewhere in between, but you could have found evidence of one of our errors at the registration desk, where extra copies of the portfolio were on sale at cost.

Trimble's Laws

1. The difficulties experienced by a woman trying to do astronomy are compounded of the difficulties experienced by anybody trying to do astronomy and the difficulties experienced by a woman trying to do anything. (A small prize will be given to the first 10 people who correctly identify the aphorism and aphorist paraphrased here.)

2. For someone interested in an astronomical career, being female has its advantages and disadvantages. They are about equal.

3. The world seems to treat some people (including me) much better than they deserve and others (perhaps including you) much worse. This is not strongly correlated with either gender or astronomical ability.

4. The only point on which Cecilia Payne Gaposchkin was ever wrong was in saying that a woman should do astronomy only if nothing else would satisfy her, because nothing else is what she would get. You also get to visit exotic places (Socorro NM, Greenbelt MD, Pasadena CA) and meet interesting men (examples deleted—Ed.).

5. Never agree to do anything for which the first sentence of the invitation mentions "women in astronomy," "female scientists," or "role models." Well, hardly ever.

STAR GAZERS

Martin J. Rees

Ninety percent of the matter in galaxies is dark and so not easily studied. Much the same could be said of Professor Martin J. Rees of Cambridge University. Well-illuminated—and illuminating—though many of his accomplishments are, a large number, especially in encouraging the work of younger colleagues, have remained relatively unknown.

Dr. Rees was born 23 June 1942, the son of a Welsh schoolmaster, and as a child he had the run of the large building and garden that housed the school. He claims not to have been particularly scientifically precocious; his first loves were large dogs and small boats, not stars and galaxies.

A 1960 graduate of Shrewsbury School, Rees received his B.A. in mathematics from Cambridge University in 1963 and his Ph.D. in 1967, working with Dennis Sciama on testing cosmological models and on properties of quasars. He has been associated with Cambridge almost continuously since then, apart from a year (1972–73) as a professor at Sussex and visiting appointments at Harvard, Princeton, and Caltech. He was one of the founding staff members of the Institute of Theoretical Astronomy, then directed by Fred Hoyle, from 1967–72, and returned in 1974 to the restructured and renamed Institute of Astronomy as Hoyle's successor in the Plumian Chair of Astronomy and Experimental Philosophy. Rees has served as director of the Institute from 1977 to 1982 and 1987 to 1992, in alternation with Donald Lynden-Bell.

Among Rees's best-known and most widely cited scientific contributions are the prediction of superluminal motion in extragalactic radio sources and the idea of powering extended radio sources via beams of low-frequency electromagnetic radiation. This latter spawned a model for pulsar-driven supernova remnants and gave rise to the now generally accepted "twin exhaust," jet, or beam models for active galactic nuclei and to the ion-supported torus mechanism for channeling jets. In recent years, in company with students and postdocs, he has focused on problems of galaxy formation, particularly though not exclusively in the framework of biased cold dark matter.

No such brief list can do justice to the scope and breadth of Rees's publications. His name appears on more than 290 "real" papers (excluding conference abstracts, book reviews, popularizations, etc.) since 1965, on topics from accreting plasmas to anthropic principles. These accomplishments have been recognized by memberships in the Royal Society, the U.S. National Academy of Sciences, and the American Academy of Arts and Sciences, as well as more than 30 prizes and honorary lectureships, including the first Bappu Memorial Award of the Indian Academy of Sciences, the Robertson Memorial Lectureship (U.S. National Academy), and the Bakerian Lectureship (Royal Society).

Less widely known are Rees's contributions as mentor and encourager. Twenty-one students have completed Ph.D.s under his direction, including Roger Blandford, James E. Pringle, Mitchell Begelman, Craig Hogan, and E. Sterl Phinney. Many of these scientific offspring now have "doctor children" of their own, and there will soon be a significant number of Rees great-grandchildren exploring the universe. In a less formal way, Rees turns up in enormous numbers of acknowledgement paragraphs and lists of referees and references for younger colleagues.

Virtually everyone who talks with Martin Rees, even casually at a conference, comes away feeling that he has made a new scientific friend and ally. Not long ago, in conversation over a coffee break, there arose the topic of which contributions various scientists were proudest of (it is not always the ones that the world remembers). When asked which of his works he was proudest of, Martin said, "Well, I hope I haven't written it yet." We think he may be right, but the mind boggles.

The cosmogonic problems are all so closely connected that we cannot expect to solve any one of them until the whole picture comes into focus. This slightly depressing view of galaxy formation and the early universe was offered toward the end of a talk given once by Rees. But do not despair; the remark came at the end of an hour of insightful exposition of just what those problems are and what branches of physics and astronomy are likely to be able to contribute to their solution.

Rees expressed many of the questions in two forms, once as a "why" and once as a "what" or a "how." The second form always made the problem sound less intractable. Why do galaxies exist at all, when, if they did not, no theorist would feel compelled to invent them? This seems almost unanswerable. But rephrase it as: what physical processes have determined the masses and sizes of the galaxies we see and what spectrum of density perturbations must have come out of the early universe to permit these processes to occur, and you have much less exotic-sounding questions, but also much less daunting ones. The existence and properties of stars are set by a balance between the forces of gravity and electromagnetism. Similarly, at least upper limits to the mass and size of galaxies come from a balance between the time scales of gravitational collapse and cooling of gas spheres.

As for the perturbations, Rees made it clear that many different patterns could have evolved into the present structure. The chief difficulty is for them to do so without introducing more temperature variation than we see into the 3 K microwave, black-body radiation, which last interacted with ordinary matter just as the perturbations were starting to grow.

Another of the major questions is the why and wherefore of dark matter. Ask it in the form, what kinds of mass energy could account for galaxy rotation curves, cluster velocity dispersions, and such without radiating much, and you are left with an embarrassment of answers. These range from small stars and large black holes to neutrinos, axions, and weakly interacting massive particles. But if you ask how dark matter can behave during the epoch of galaxy formation, then the plethora of possibilities collapses to three main types—hot, cold, or baryonic dark matter. Rees claims on this issue to be an agnostic, attaching roughly equal probability to each of the major candidates, but reserving 25 percent for something we haven't thought of yet. Laboratory experiments currently under way to look for WIMPs and axions could redistribute this probability quickly. In the meantime, properties of dark matter are most severely constrained by its having participated in the process of galaxy formation.

The third question concerned quasars: Why do they exist, or alternatively, what is happening at the centers of a few galaxies (or many galaxies, but briefly) to make them flare up? We awaited with trepidation a judgment on what connection this process might have with the epoch of galaxy formation. We are still waiting. The connection is apparently not simple. But the mere existence of quasars before a redshift of four pushes the epoch of galaxy formation back farther than some models prefer.

Undoubtedly, we do not yet have a complete picture of the formation and evolution of structure in the early universe, but some edges are beginning to emerge from the hypo, and we can already see Martin Rees in the front row, grinning.

Maarten Schmidt

M aarten Schmidt discovered quasars in 1963. This is quite often the only thing elementary astronomy students learn about Schmidt or, for that matter, about quasars. We know better!

Schmidt is a native of Groningen, the Netherlands, where he was born on 28 December 1929, and a graduate of Leiden University, where he earned a Ph.D. in 1956. His Dutch colleagues and friends still call him Maart, but you should attempt this only if you have a good Dutch "r" in your repertoire of phonemes.

Since 1959, Schmidt has been permanently associated with the California Institute of Technology, where he is now the Francis L. Moseley Professor of Astronomy. Along the way, he served terms as Executive Officer for Astronomy and Chairman of the Division of Physics, Mathematics, and Astronomy at Caltech, Director of the Hale Observatories, and President of the American Astronomical Society.

Schmidt has also along the way garnered an impressive array of awards and honors. These include the Warner Prize and Russell Lectureship of the AAS, the Halley Lectureship and Gold Medal of the Royal Astronomical Society, the Schwarzschild Lectureship of the Astronomische Gesellschaft, honorary degrees from Yale and Wesleyan universities, and memberships in the U.S. National Academy of Sciences (as Foreign Associate) and Koninklijke Nederlandse Akademie van Wetenschappen (as Correspondent). Through all this, he has somehow managed to keep not only his sense of humor but also an affection for bow ties and most of his hair.

Schmidt's career as an observational astronomer began at the age of thirteen, with the spotting of a double star in Lyra through a homemade Galilean telescope. This result does not seem to have been published anywhere, but his first widely cited paper appeared less than ten years later, in 1951. It reported work with his advisor, Jan Oort, indicating that comets age very quickly, virgin ones having noticeably stronger dust tails than do others. Schmidt soon turned from optical studies to the infant discipline of radio astronomy, demonstrating the density enhancements associated with coronal streamers through their effect on the polarization of background radio sources.

This involvement with radio astronomy led to the work on star formation rates and galactic structure and evolution for which Schmidt was best known in pre-quasar days. His 1965 model of the galactic rotation curve and gravitational potential remains a standard to which more recent models are compared. It was built on detailed studies of galactic 21-centimeter radio emission and the galactic properties implied thereby as well as upon careful consideration of the kinematics of globular clusters and Cepheid variables. In the field of galactic evolution, Schmidt's conclusion that star formation rates scale with the square of gas density is both frequently cited and frequently confirmed. And he was among the very first to understand the importance of what is now called the G dwarf problem—the fact that the solar neighborhood has far fewer metal-poor stars than it would if star formation and nucleosynthesis had proceeded hand in hand over the age of the galaxy. Attempts to resolve this discrepancy still dominate discussions of galactic chemical evolution.

The discovery of quasars in 1963 occurred as part of a relatively new project, Schmidt's continuation of Rudolph Minkowski's program to identify radio sources. Inevitably, the direction of his research changed, though not the impressive quantity or quality. The years since 1963 have seen an average of 5.7 papers per year carrying his name as author or co-author. The most widely cited of these (*Astrophysical Journal* **151,** 393 [1968]) reported a new method for doing statistics on objects in samples with complicated completeness limits. Applying the method, called V/V_{max}, to quasars, Schmidt showed that they occur preferentially at large distances, and so were commoner in the past. An application to halo stars (*Astrophysical Journal* **202,** 22 [1975]) showed that the method could also be used to derive luminosity functions from incomplete data. Its power, especially for investigating the faint end of the stellar luminosity function, is only just beginning to be appreciated.

A quasar is not forever, according to Schmidt. Not only do the individual objects have lifetimes less than one percent of the 10- to 20-billion-year age of our universe but, in addition, the era during which most of them brightened, flourished, and faded away was also short, perhaps 10 percent of the age of the universe. This brief period, he suggests, must have been a significant epoch in cosmic history.

Serendipity clearly played a role in the discovery of quasars, which occurred between 1960 and 1963, though not, we suspect, so large a one as he modestly suggests. Perhaps more important (and more important for us as a model) was the close collaboration with radio astronomers Cyril Hazard and Thomas Matthews and spectroscopists Jesse Greenstein and J. Beverley Oke. This made possible the efficient identification of strong radio sources and the elimination of many possible interpretations of their emission lines that led to Schmidt's attempt to construct an energy-level diagram for a hypothetical atom and so to his recognition of the redshifted hydrogen lines in the spectrum of 3C 273.

Schmidt has often emphasized the large numbers of people who have contributed to our present understanding of quasars. But most of the defining properties were already evident in his initial work—star-like images (hence the name *quasi-stellar* and its various permutations), rapid variability (implying small size), and broad emission lines with large redshifts.

Schmidt and his collaborators quickly decided that these were best interpreted as cosmological red shifts, placing the quasars at large distances and their heyday in the remote past. Most astronomers agree, and Schmidt, among others, has suggested that the issue might never have become controversial if quasars had been discovered after rather than before we recognized the rapid variability of Seyfert nuclei. In any case, the only alternative to placing the quasi-stellar objects at cosmological distances (so that they probe an early phase of cosmic evolution) is to accept that we live in a very special place, near the center of a hole in the quasar population.

Schmidt plots of the number of quasars as a function of age of the universe (or cosmic time) in linear rather than logarithmic coordinates. A steep spike at three billion years (for $H = 70$ km/sec/Mpc and small q_o) represents a density 130 times the present one. The paucity of quasars at earlier times is also significant both in the data collected by Schneider, Gunn, and Schmidt at Palomar and in that from Paul Hewitt's group in Cambridge, England.

The sharp maximum must signify an extraordinary event in the early universe. Astronomers have been looking for such an epoch—that of galaxy formation—for a generation. Schmidt suspects that these two special epochs may be intimately related and claims to await with some trepidation theorists' conclusions on the point.

It is safe to say that the rise of the quasars was just one point along the curve representing the rise of astronomer Maarten Schmidt, and that no fall is in sight for either the discovery or the discoverer.

Beatrice M. Tinsley

In a career that spanned less than 14 years, from Ph.D. dissertation in 1967 to her death in 1981, Beatrice Muriel Hill Tinsley published some 114 papers, many of which continue to be cited heavily. She created, as nearly single-handedly as anything in modern astrophysics can be said to be single-handed, the discipline of modeling the evolution of galaxies to account for their chemical compositions, luminosities, colors, and gas contents as a function of time and applied those models to fundamental astronomical problems, including that of determining the age and geometry of the universe as a whole.

In her years as a professor at Yale, Tinsley was formal mentor to half a dozen graduate students and informal guide and counselor to many more, as well as a valued colleague and collaborator of astronomers all over the world. All this occurred against a background of an early career at somewhat inappropriate and not entirely hospitable institutions, a mid-career shared with the raising of two small children, and three final years darkened by the shadow of terminal melanoma.

After her death, her father, Edward Hill, compiled and circulated among friends and colleagues a brief biography, *My Daughter Beatrice,* published by the American Physical Society in 1986. A large fraction of the text consists of passages from letters Beatrice wrote to her family, from her school years onward. In the extracts that follow, her words are set off in double quotes, her father's words in single quotes. My bridging and explanatory notes appear unadorned or, when they intrude on the extracts, within brackets.

Some of the experiences Tinsley relates will be recognized and shared by virtually all scientists—the struggle, for instance, to explain to one's family just what it is one is trying to do. Others, like finding oneself the only girl in a lecture course of 250, are the common property at least of female scientists. But, obviously, much of what Beatrice and her father recorded were personal events, not universal attributes or even prototypical ones of science or women in science. Growing up the only child of a chemist father and a mother with a flare for language, within easy driving distance of both UCLA and Caltech, I found the path from one educational institution to another—college, graduate school, postdoctoral fellowships, tenure track, and eventually tenured job doing astrophysics—very much the path of least resistance. Other scientists will have experienced something in between, or something entirely different. And, mercifully, very few of us indeed are forced to choose between family and a suitable job in quite so harsh a way; still fewer compelled to face so devastating and premature an ending to our scientific endeavors.

But Beatrice Tinsley was a wise and wonderful person, as well as an unusually productive scientist; thus, her experiences and reactions to them cannot help but carry lessons for the rest of us.

Family Background

'From my side of the family, Beatrice gained, as I have already stated, some of her physical appearance, though her fine bone structure was more that of a Morton. Her musical ability could be traced to both her mother and mine, who had been educated to be a professional pianist and kept up her playing to near-professional standard throughout almost all of her life. Jean [Morton Hill] was a cellist, as well as being a good accompanist, but her principal interest was in writing. She had more than one play performed, and published three novels as well as numerous short stories. I had been a scholar at Oxford and later spoke a good deal in public in addition to writing. So I like to think that Beatrice's gift for expressing complex ideas in simple terms came in some degree from both parents. It was one of her great talents and was responsible, I have no doubt, for her being in such demand in the last years of her life as a speaker and a lecturer in Europe, America, and even farther afield.'

Childhood and Schooling

Beatrice Tinsley was born in Chester, England, 27 January 1941. 'We arrived in Christchurch [New Zealand] at the end of August 1946. . . . Beatrice was placed in kindergarten and the report for her second term there survives: Her number work is excellent, and her printing neat, it stated.'

'One day [in 1953] Jean drew my attention to a timetable stuck to the wall above Beatrice's bed. It was divided in squares for each hour of the week. Those not to be spent at school were carefully allotted to music practice or play or homework, and this was the first of many such timetables, which were adhered to with discipline.'

'Beatrice's high school reports, of which I still have the majority, make almost monotonous reading. She gained consistently high marks in all subjects and won her form prize each year! . . . In the third year she won cups for Latin, mathematics, and speech, and every one of her subjects was described as good or very good indeed.'

'Looking back I can see that she displayed considerable character not only in her determination to do mathematics for [a] scholarship, but in the disciplined way she had concentrated on all her studies during all her high school years.'

A summer romance led to an engagement, which, however, faded away as she entered 'enthusiastically into student life at Canterbury University.' In retrospect, Beatrice did not have all the traits folklore associates with great achievement: she was the middle of three daughters, neither first, last, nor only, and the family background and finances were such that there was never any question that the daughters could all attend college if they wanted to. On the other hand, she was unequivocally precocious in the use of words and numbers, as well as in music; reading, writing, calculating things, and violin playing being important parts of her life from childhood onward.

Undergraduate Years

'A few days after the start of her first term at Canterbury, Beatrice explained why she was already finding university study such an enormous stimulus . . . she wrote, "I'm meant to be attending to a bloke who's lecturing on the trig we've all known for two years, so I'm writing this to save myself from going to sleep. . . . Physics is really super. . . . Wonderful training in taking nothing for granted. It's our job to try to prove a theory wrong rather than assuming it's right. Everything must be thought of from a logical and perfectly basic beginning . . . we're constantly reminded that in science a theory is accepted only as long as it isn't disproved by experimental evidence. I doubt people would mind what theories or questions anyone liked to put forward here if they felt like it. It gives one the

courage to think originally, and that is the beginning of research. Of
course I've known in theory for a long time how all this is, but marvellous
to be in a position to think and experiment *myself*. . . . Why *is* male company so much better than endless females? Perhaps a reaction to living
here. In all our lecture classes, which are well over a hundred, there are no
more than ten girls! Very nice state of affairs—for us that is."

By the second year, choices had to be made—"was I interested in theoretical or practical physics, and the answer's theoretical, which is of
course applied Maths and not Physics. . . . The trouble with me right now
is I want *both:* the maths to reason with and the physics to apply it to.
Consequently I've left it open by enrolling for Pure and Applied Maths
Two and Physics Two, without tying myself to the honors course in either.
I've put physics as my major subject . . . Mary Veitch is also doing
physics two, so we'll have a wonderful time—That physics dept. has been
exclusively male for years now."

"I've spent £15 on books I and haven't got the lot yet! . . . I saw in the
Varsity book shop this gorgeous book for 58s/6d and I *had* to get it. It's an
American book and therefore beautiful, called *Theories of the Universe,*
and is a collection of writings from Plato to Einstein and Bondi, wonder-
fully put together and arranged, and being historical it won't date. One
could more or less add to it oneself."

'With her [second year] exam results all A's with marks in the nineties,
she set herself to become a cosmologist. Evidently I did not know what
the word meant, for she wrote at the start of the [third] year's work . . .'
"Really, it's theoretical astronomy; studies of theories about the universe,
its origin, structure etc. That's all! Relativity theory is absolutely basic to
it. . . . I gather it is inclined to be so mathematical it gets away from reality; apparently why it's necessary for me to keep up with Physics. It's hard
to say what I might actually end up studying, because when things get
moving in Cosmology, the whole aspect of the subject seems to change
overnight. . . . The fascinating thing about theoretical Physics is that you
can never learn about it fast enough because there's always more being
discovered to learn! And you have to specialize as soon as possible, otherwise you'll never get to the frontier of any branch of knowledge and so be
able to have some new ideas."

'Her decision to become a cosmologist gave her an even greater incen-
tive to work.' "Now I feel everything I learn is really getting me somewhere, and all useful; it's different to have a *definite* aim ahead, instead
of vaguely wanting to do research and not knowing *what.* " 'Beatrice
once more gained A's for all her exams for B.Sc.'

From Bachelor's Degree to
Graduate School and Motherhood

'By the beginning of October [1960] she . . . became engaged to Brian Tinsley.' "I didn't think extra comments would be necessary with that photo. Behold two happy people." 'Back in Christchurch for a new term, Beatrice decided on three subjects in Physics for her M.Sc., though her letters home were not unnaturally filled with preparations for her wedding [in May 1961].'

'She had forecast that her M.Sc. thesis would probably not be on anything' "very astronomical, because it's not studied here." 'As it turned out she chose what seems to me at least an obscure topic in Physics, theory of the crystal field of neodymium magnesium nitrate. For this she gained first class honors, but, as soon as the thesis was complete, was agog to return to cosmology.'

'Not till after Beatrice's death, when Professor Weybourne kindly sent me a print-out of all her results at Canterbury University, did I learn that she had been awarded the Haydon Prize for Physics and The Charles Cook, Warwich House, Memorial Scholarship. Another member of the staff wrote at the same time, expressing the almost incredible, if very pleasing, opinion that she had been perhaps the most brilliant student who ever attended the university.'

Her first real job was teaching science in a girls' high school. "I've really enjoyed teaching this year—it's a great thing to try helping young girls on the way to the world and they are such a nice crowd. We had some awfully worthwhile discussions in the last week, as I read them some interesting books etc. on the nature of science and the job of the scientist, his responsibility to society and so on."

"Big talks about possible futures, as I'm expected to hand in applications for overseas scholarships. . . . I can but refuse if I'm offered one to go where Brian isn't. . . . . Another place has arisen to which Brian and I are putting out feelers, the Southwest Center for Advanced Studies, in Dallas, Texas, . . . which is setting up a big new lab for Earth and Planetary Sciences and would apparently welcome people like Brian" [whose research in planetary atmospheres, aurorae, and the like is still underway at the University of Texas, Dallas].

In July 1963 'the Grants committee . . . agreed to her holding her scholarship at the Southwest Center.' And by December they had together attended one of the early Texas Symposia on Relativistic Astrophysics; "been inconspicuous and understood varying amounts of papers and discussions. The atmosphere has been really thrilling. . . . Just to hear and see the varied interesting personalities of so many great scientists was marvellous. The pooling of ideas must have been of tremendous value to researchers, but I don't think they have any more certainty as to what the strange objects [the newly-discovered quasars] are than they did [before]. I wish I knew enough to appreciate it fully."

"Big changes afoot in my life. In September [1964] I am enrolling in the Astronomy Department of the University of Texas in Austin for a Ph.D. . . . Here I seemed to have reached rather a dead end. Apart from summer visitors, the line of interest among the Division was very much General Relativity as a mathematical game, with applications an incidental sideline if at all . . . but the U of T in Austin has a well known Astronomy Dept., headed by an exceptionally brilliant . . . young man, and owning the 4th largest telescope in the world. . . . I am the only woman student there and the others seem a nice crowd."

Tinsley quickly completed the requisite coursework and qualifying exam and settled down to a thesis comparing observed properties of galaxies with the predictions of various cosmological models. 'She stated that she had little idea how much was needed to make a Ph.D dissertation but hoped to finish the degree up in a year or two and' "ultimately (i.e., when we are at the children-grown-up stage) look for a University teaching + research post." 'Her reference to children . . . was explained by their intention to adopt a baby, due to be born in New Zealand in August. In spite of much medical advice and even a small operation . . . there was no sign of Beatrice having a child of her own.'

Meanwhile, her thesis advisor was "very pleased with all my results. . . . It seems I've done enough research to get a Ph.D. with and now have only to write it up. . . . That's only a few months work, even with a baby," 'and, by December 17th [1966] Beatrice had her Ph.D. She hadn't felt very proud of herself after the oral,' "but they passed me. Afterwards I felt very anti-climactic and deflated, seeing no purpose ahead but washing and sweeping. But by now I am altogether happy at the thought of doing astrophysics in the spare time left by Alan and Brian, and being Alan's mother is my No. 1 priority as long as he is totally dependent all day. I might consider applying for a post-doctoral fellowship to do research at home for the next few months, and then spend a lot of it on getting daily help to do all the housework, so I can study while Alan is asleep and in the evenings, most of which Brian spends at the Center anyway." In 1968, she and Brian applied "to adopt a little girl [Terry] from a Dallas adoption agency—I'm terribly excited about it."

The Postdoctoral Years

'In early March [1968] her paper, incorporating her thesis work, appeared in the *Astrophysical Journal*'—"very prestigious, very slow. . . . What makes me happier still is that people have referred to my work and found it useful."

'With two small children, plus a husband and house to care for, Beatrice necessarily became very domesticated. In July she managed to spend an evening in the library reading the latest science journals and rather sadly finding' "how different the current problems will be by the time I'm active again." 'She was the more tied to the house by Brian being so often away on his work.' A delayed reprint request in February 1969 made her realize "how long it is since I did any astrophysics! Time enough ahead—life is very full as it is."

UT Dallas provided a title of visiting scientist and some money for page charges and computing, but in 1970 'Brian was in Brazil for two months, arriving home on December 12th, only just in time for Beatrice to do his washing and ironing before she flew to Austin for a conference on Relativistic Astrophysics, where there were' "lots of useful conversations with people whose names are familiar from their research papers. Also, which is important to me in my relative isolation from astronomers, a number of people were very generous with pre-publication copies of exciting new data."

In 1971, the National Science Foundation came through with a research grant, in time to pay for a trip to the August American Astronomical Society meeting in Amherst, Massachusetts. 'She was impressed when a' "very eminent woman astronomer—great enough to win honors in any contest—turned down a prestigious prize offered to women astronomers (only) on the ground that special honors and discrimination for women should be abolished. . . . It certainly woke up some of those who were there as to what problems (not all discriminatory of course) women face and how deeply they are felt! I've written (I hope politely) to the committee."

In early 1972, children in tow, Beatrice took off on the first of many brief visiting scientist appointments, this one at Caltech. The experience was positive: "I'm working very hard, enormously enjoying life being surrounded by eminent astronomers and astrophysicists. . . . I'll be going back to Dallas with enough ideas to last for years, not to mention many good new friends. . . . I'd love to learn to observe when the kids are older. The vast scale of the 200-inch and its dome is terribly impressive."

'Back in Dallas,' Beatrice worked "very hard" on [a] "proposal for an Astronomy Dept. at U.T.D. It looks hopeful, fairly sure to get approval here, but the State Coordinating Board may not want it. The proposal has to be written in the most nauseating jargon. I read three other proposals through, drank a glass of wine, then translated my straightforward English into a fairly good imitation."

'From another visiting position, at the University of Maryland in spring 1973, she wrote,' "I find myself now an accepted member of the community of cosmologists and astrophysicists—gratifying fulfillment of long dreams! The work is a great pleasure to me." 'From Caltech, she reported the satisfaction of going' "back there among all those famous astronomers and [being] greeted like and treated as a respected colleague"; and from Cambridge, England, "learning a lot from others, and getting a lot of research done with people from all over the world who gather in Cambridge for the summer." She received the 1974 Annie J. Cannon Prize as the outstanding female astronomer under 35.

From Texas to Yale

'I do not propose to record all the reasons she gave then and later for the divorce, but undoubtedly conflict between their careers was a contributing factor.' Trying to go on working at U.T.D., she wrote, "had reduced me to a mental state of anguish. Hard to explain! I am a good scientist, and among my peers treated like a full and respectable person and feel of *worth*. U.T.D. has kept me at the nearest possible level to nothing and there is *no one* who knows enough about astronomy to care in the least for my work. . . . To be rejected and undervalued intellectually is a *gut* problem to me, and I've lived with it most of the time we've been here, apart from extended visits to Caltech and Maryland and shorter trips to meetings and so on." Job offers from Bell Labs, Chicago, and Yale were quickly forthcoming.

'In what had presumably been a last effort to save her marriage she had applied to be made head of the Astronomy Department she had designed [at UT Dallas]. There had been no reply and at [a] party the man who should have answered remarked casually, I have a letter from you, don't I, that I must answer some time. She told me later that she took a somewhat vicious pleasure in replying,' "You needn't bother now. I'm choosing between Chicago and Yale." The children remained in Dallas, "in the place they have always known, with their friends, school, child minders, etc.," and she had no doubt that "Brian is an excellent parent." They visited her often during vacations, sharing trips to conferences in England and elsewhere.

Beatrice chose Yale. Her career there, from associate to full professor in three years, an Alfred P. Sloan Foundation Fellowship, and so forth, differed from most academic careers only in its enormous productivity and tragic brevity. En route to Yale, she wrote from a conference in Venice (in summer 1975) that she was "very moved to be giving a review of cosmology at Hoyle's sixtieth birthday, since it was his books that I read at high school that introduced me to the subject." Less than six years later came her last birthday greetings to her father and stepmother: "I think of you a whole lot, not only on birthdays, and wish you strength and happiness in coming days. I honestly don't think that length of life is important. Very much love from Beatrice."

Fritz Zwicky

F ritz Zwicky was born on 14 February 1898 in Varna, Bulgaria, the eldest child of a Swiss merchant father and a Czech mother. Although most of his career was spent in the United States, he remained throughout a citizen and voter of the Swiss canton of Glarus. His first and second degrees, earned at the Eidgenoessische Technische Hochschule in Zürich in 1920 and 1922, were for investigations in what would now be called condensed matter physics, though the work for which he is best remembered was in astrophysics and cosmology. He also made significant contributions during and after World War II to rocketry and jet propulsion, for which he received the 1949 Medal of Freedom from President Truman.

These clashes of both national and scientific cultures were undoubtedly major contributors to Zwicky's unique way of viewing the universe. He was (only half jokingly) described as speaking seven languages, all badly, and was much given to drawing (sometimes imperfect) analogies between widely diverse phenomena. A good deal of secondary literature characterizes Zwicky as opinionated, cantankerous, and otherwise generally difficult to get along with, but these traits seem to have been strongly situation dependent and were balanced by a deep devotion to family and close friends, serious involvement in charitable endeavors (including the rebuilding of scientific libraries destroyed during World War II and the care of children orphaned thereby), and a sort of old-fashioned chivalry toward those he perceived as relatively defenseless.

Zwicky's most cited book, the 1957 *Morphological Astronomy,* as well as many of his scientific papers and lectures, show that he was typically more concerned with making sure that all possible phenomena and processes were considered than with hitting upon the dominant ("right") one in a particular context. Thus the tendency among astronomers today to focus exclusively on the questions to which he clearly had the first right answer is, in its way, just as misleading as his contemporaries' tendency to focus on the areas where what he said was demonstrably wrong, implausible, or simply out of fashion.

To take some of the latter cases first, Zwicky took fairly seriously both tired light and gravitational redshifts as alternatives to cosmological expansion to explain the large redshifts of distant galaxies (but he systematically used redshift as a reliable distance indicator in his own work). He frequently and firmly denied that clusters of galaxies are, in turn, clustered—though many of the structures that he called multicored clusters would today be called superclusters, and his value for the size of a characteristic cluster cell, 40 Mpc (for $H = 100$ km/sec/Mpc), is nearly a modern supercluster. The pixielike grin with which he sometimes concluded the calculation that turns the absence of structure larger than 3×10^{25} cm into a value of 10^{-63} grams for the mass of the graviton may have meant that here, too, he was merely exploring the possibilities. His "pygmy stars" and "object hades" (stable configurations more compact than neutron stars) were clearly an example of this, and if Zwicky had lived into the era of strange quark stars (now an almost respectable alternative to the highest-density neutron stars) he would surely have claimed them as belonging in one of his previously empty morphological boxes.

Subjects where Zwicky is generally regarded as being first on the field with a major piece of the right answer include supernovae, existence of large quantities of dark matter in rich clusters of galaxies, and gravitational lensing of galaxies.

The possibility of two classes of novae, with different peak brightnesses, had been advanced in the 1920s by H. D. Curtis and Knut Lundmark (a sort of Danish Zwicky, who was also the first to plot red shifts of galaxies versus a distance indicator and find a systematic trend, now called Hubble's law). But a pair of 1934 papers by Walter Baade and Fritz Zwicky first clearly defined the category of supernovae, with a handful of examples, including the 1054 event whose remnant is the Crab Nebula. In addition, they proposed that supernovae were the main energy source for cosmic rays, and that the energy source for supernovae was, in turn, the conversion of a normal star into a neutron star. These are essentially the modern views, for at least some supernovae and some cosmic rays.

The intellectual history of neutron stars is currently in some disorder. The 1932 discovery of the neutron by Chadwick and the 1931 proposal by Milne that novae result from the conversion of a normal star to a white dwarf are documented in the archival journals. But the belief that Lev Landau thought of neutron stars very shortly after Chadwick's discovery rested on an after-dinner talk by Leon Rosenfeld, more than 40 years after the event, and falls afoul of records at the Niels Bohr Institute in Copenhagen which show that Landau had already returned to Russia at the time mentioned by Rosenfeld. It is possible, therefore, that Baade and Zwicky were the inventors of the concept of neutron stars as well as of supernovae and of the association.

Baade's interest returned to galaxies and stellar populations, but Zwicky engaged actively in searches for supernovae and interpretation of the data both before and after the war. He himself found 122 SNe, a still unequaled record; many of the most productive recent searches are closely modeled on his programs with the 18-inch and 48-inch Schmidt telescopes at Mt. Palomar. Zwicky early connected up supernovae with cosmology by suggesting (simultaneously with Olin Wilson in 1939) that they could be used as reliable distance indicators out to very large distances. This has since been attempted in several different ways, and the method is only just coming into its own.

The discovery of dynamically important dark matter came with the measurements of redshifts for about a dozen individual galaxies in the Coma cluster (first published in 1933). Zwicky pointed out that gravitational binding of galaxies with the observed velocity dispersion would require a mass more than 100 times that implied by the amount of light coming from the galaxies themselves. He specifically proposed (at various times) gas, faint galaxies, and intergalactic stars as the dominant constituent of the rich clusters. A similarly large velocity dispersion was found for the Virgo cluster by Sinclair Smith in 1936, and he proposed similar explanations. We now believe their data for the existence of dark matter, but not their candidates for its nature.

Finally, in 1937 Zwicky was the first to propose the one kind of gravitational lens that we now think has been seen—the sort where the appearance of a galaxy is strongly distorted by the bending of its light rays as they pass another galaxy en route to us. The history of the idea of gravitational lensing is again complex. The first published suggestion that one star might lens another seems to have been that of O. Chwolson of Petrograd in 1924. But a Zwicky footnote mentions that E. B. Frost of Yerkes Observatory had a plan to look for the effect as early as 1923. Einstein discussed lensing of stars in 1936, saying that he had been asked to do the calculations by R. W. Mandl. And Zwicky says he had the idea from V. K. Zworykin (better known for contributions to the technology of television) who, in turn, had it from Mandl. Zwicky's specific contributions were to show that lensing of galaxies was much more probable than lensing of stars and to indicate that the observed properties of the lensed and lensing objects would provide independent estimates of the masses of galaxies. He obviously expected support for the large masses implied by his work on Coma, not the small ones associated with the luminous parts of galaxies. This has, indeed, happened, though it has taken some 50 years to see that Zwicky was right.

Zwicky's death on 8 February 1974 frustrated his often-stated ambition to be the first astronomer to live in three centuries.

Further Readings

W. Baade and F. Zwicky, "On super-novae" and "Cosmic rays from supernovae," Proc. Nat'l. Acad. Sci. **20,** 254 and 259 (1934).

O. Chwolson, "Über eine mögliche Form fiktiver Doppelsterne," Astron. Nachr. **221,** 329 (1924).

A. Einstein, "Lens-like action of a star by the deviation of light in the gravitational field," Science **84,** 506 (1936).

K. G. Hufbauer, "Fritz Zwicky," in *Dictionary of Scientific Biography,* Supplement (1990).

S. Smith, Astrophys. J. **83,** 23 (1936) (mass of the Virgo cluster).

O. C. Wilson, Astrophys. J. **90,** 634 (1939) (supernovae as distance indicators).

F. Zwicky, "On the redshift of spectral lines through interstellar space," Proc. Natl. Acad. Sci. **15,** 773 (1929).

F. Zwicky, "Rotverschiebung extragalaktischer Nebel," Helv. Phys. Acta **6,** 110 (1933).

F. Zwicky, Phys. Rev. **55,** 726 (1939) (supernovae as distance indicators).

F. Zwicky, "On the masses of nebulae clusters," Astrophys. J. **86,** 217 (1937).

F. Zwicky, *Morphological Astronomy* (Berlin: Springer-Verlag, 1957).

REFERENCES

References

Cheops's Pyramid *(page 3)*

1. W. Fl. Petrie, *The Pyramids and Temples of Gizeh* (1883), p. 53.
2. W. S. Smith, *The Art and Architecture of Ancient Egypt* (New York: Penguin Books, 1958), p. 30.
3. Alexander Badawy, *A History of Egyptian Architecture I* (Giza, 1954), p. 163.
4. Robert Baker, *Astronomy* (New York, 1950), p. 57.
5. Otto Neugebauer and Richard A. Parker, *Egyptian Astronomical Texts I: The Early Decans* (London, 1960), p. 25.
6. *Ibid.,* p. 110.
7. S. R. K. Glanville, *The Legacy of Egypt* (Oxford, 1942), pl. 32.
8. Paul V. Neugebauer, *Tafeln zur astronomischen Chronologie I: Sterntafln* (Leipzig, 1912), pp. 8 and 20.
9. *Ibid.,* pp. 21–82.

Star of Bethlehem *(page 9)*

1. David Hughes, *The Star of Bethlehem: An Astronomer's Confirmation* (New York: Walker, 1979).
2. J. Finegan, *Handbook of Biblical Chronolgy* (Princeton, N. J.: Princeton University Press, 1964).
3. J. L. McKenzie, *Dictionary of the Bible* (New York: Macmillan, 1965), p. 534.
4. D. H. Clark, J. H. Parkinson, and F. R. Stephenson, Q. J. R. Astron. Soc. **18,** 443 (1977).
5. T. D. Barnes, J. Theological Studies **19,** 204 (1977).
6. Clark *et al.,* note 4; A. J. Morehouse, J. R. Astron. Soc. of Canada **72,** 65 (1978).
7. Clark *et al.,* note 6.
8. Morehouse, note 6.
9. F. Whipple, personal communication, 1980.

In Their Own Words *(page 20)*

1. J. C. Kapteyn, Astrophys. J. **55**, 302 (1922).

2. J. H. Jeans, Mon. Not. R. Astron. Soc. **82**, 130 (1922).

3. J. H. Oort, Bull. Astron. Instit. Neth. **6**, 249 (1932).

4. H. W. Babcock, Lick Obs. Bull. **19**, (No. 498), 41 (1939).

5. H. W. Babcock, private communication, 1988.

6. E. Holmberg, Lund Ann. **6**, 102 (1937).

7. F. Zwicky, Helv. Phys. Acta **6**, 110 (1933).

8. S. Smith, Astrophys. J. **83**, 23 (1936).

9. G. O. Abell, Astron. J. **66**, 580 (1961).

10. S. von Hoerner, Astron. J. **66**, 580 (1961).

11. A. Finzi, Mon. Not. R. Astron. Soc. **127**, 21 (1963).

12. J. P. Ostriker, P. J. E. Peebles, and A. Yahil, Astrophys. J. **193**, L1 (1974).

13. J. Einasto, A. Kaasik, and E. Saar, Nature **250**, 310 (1974).

14. S. S. Gershtein and Ya. B. Zel'dovich, JETP Lett. **4**, 174 (1966); p. 120 in Russian.

15. G. Steigman and M. S. Turner, Nucl. Phys. **B253**, 375 (1985).

16. F. Zwicky, Phys. Rev. **51**, 290 (1937).

17. *Ibid.,* 679.

18. *Ibid.,* 290.

19. A. Einstein, Science **84**, 506 (1924).

20. O. Chwolson, Astron. Nachr. **221**, 329 (1924); A. Einstein, Astron. Nachr. **221**, 329 (1924).

21. Zwicky, note 16, 679.

22. H. Shapley, Mon. Not. R. Astron. Soc. **94**, 815 (1934).

23. F. Zwicky, Astrophys. J. **86**, 218 (1937).

24. F. Zwicky and K. Rudnicki, Astrophys. J. **137**, 718 (1963).

25. *Ibid.*

26. V. C. Rubin *et al.*, Astron. J. **81**, 687 and 719 (1976).

27. V. C. Rubin, Astron. J. **56**, 47–48 (1951)

28. G. de Vaucouleurs, Astron. J. **63**, 253 (1958).

29. O. J. Eggen, , D. Lynden-Bell, and A. R. Sandage, Astrophys. J. **136**, 748 (1962).

30. J. H. Jeans, *Astronomy and Cosmogony,* 2nd ed. (Cambridge: Cambridge University Press, 1929), sec. 322.

31. E. M. Lifshitz, JETP **16**, 587 (1946); in Russian.

32. W. B. Bonnor, Mon. Not. R. Astron. Soc. **117**, 104 (1956).

33. C. F. von Weizsäcker, Z. Astrophys. **22**, 319 (1943); von Weizsäcker, Astrophys. J. **114**, 165 (1951); W. Heisenberg and von Weizsäcker, Z. Phys. **125**, 290 (1948).

34. F. Hoyle, Astrophys. J. **118**, 513 (1953).

35. G. Gamow and E. Teller, Phys. Rev. **55**, 654 (1939).

36. G. Gamow, Rev. Mod. Phys. **21**, 372 (1949).

37. G. Gamow, Proc. Natl. Acad. Sci. **40**, 480 (1954).

38. Bonnor, note 32.

39. A. D. Sakharov, JETP **49**, 345 (1966); in Russian.

40. R. A. Alpher and R. Herman, Phys. Today, August 1988, p. 24, and November 1988, p. 157

41. A. McKellar, Publ. Dominion Astrophys. Obs. **7**, 251 (1941).

42. G. Herzberg, *Spectra of Diatomic Molecules* (New York: Van Nostrand, 1950) p. 496.

43. A. A. Penzias, and R. E. Wilson, Astrophys. J. **142**, 420 (1965).

44. R. A. Alpher, H. Bethe, and G. Gamow, Phys. Rev. **73**, 803 (1948).

45. R. A. Alpher and R. Herman, Nature **162**, 775 (1948).

46. R. A. Alpher and R. Herman, Phys. Rev. **75**, 1092 (1949).

47. C. Hayashi, Prog. Theor. Phys. **5**, 224 and 235 (1950).

48. E. A. Ohm, Bell Syst. Tech. J. **40**, 1073 (1961).

49. Ya. B. Zel'dovich, JETP **43**, 1561 (1962), in Russian; Zel'dovich, Usp. Phys. Nauk **6**, 475 (1963), in translation.

50. A. G. Doroshkevich and I. D. Novikov, Dok. Akad. Nauk SSSR **154**, 809 (1964); in Russian.

51. Penzias and Wilson, note 43.

Relativistic Astrophysics *(page 28)*

1. F. Hoyle and W. A. Fowler, Nature **197**, 533 (1963).

2. C. Hazard, M. B. Mackey, and A. J. Shimmins, Nature **197**, 1037 (1963); M. Schmidt, Nature **197**, 1040 (1963); J. B. Oke, Nature **197**, 1040 (1963); J. L. Greenstein and T. A. Matthews, Nature **197**, 1041 (1963).

3. *Ibid.*

4. I. Robinson, A. Schild, and E. Schucking, eds., *Quasi-Stellar Sources and Gravitational Collapse* (Chicago: University of Chicago Press, 1965).

5. J. N. Douglas *et al.,* eds., *Quasars and High Energy Astrophysics* (New York: Gordon & Breach, 1969).

6. M. P. Ulmer, ed., *13th Texas Symposium* (Singapore: World Scientific, 1988).

7. R. Giacconi *et al.,* Phys. Rev. Lett. **9**, 439 (1962).

8. S. Bowyer *et al.,* Nature **201**, 1307 (1964); Science **146**, 912 (1964).

9. W. Israel, in S. Hawking and W. Israel, eds. *300 Years of Gravitation* (Cambridge: Cambridge University Press, 1987), p. 199.

10. W. Baade and F. Zwicky, Proc. Natl. Acad. Sci. **20,** 254 and 259 (1934).

11. W. Israel, talk at American Physical Society session marking 75th anniversary of General Relativity, 1990.

12. C. K. Seyfert, Astrophys. J. **97,** 28 (1943).

13. L. Woltjer, Astrophys. J. **130,** 38 (1959).

14. F. Hoyle, *Some Recent Researches in Solar Physics* (Cambridge: Cambridge University Press, 1949).

15. H. A. Alfvén and N. Herlofson, Phys. Rev. **78,** 616 (1950).

16. V. I. Ginzburg, Ann. Rev. Astron. Astrophys. **28,** 1 (1990).

17. P. Lenard, Ann. Phys. (Leipzeig), ser. IV, **65,** 40 (1921).

18. A. Einstein, Science **84,** 506 (1937).

19. F. Zwicky, Phys. Rev. **51,** 290 and 679 (1937).

20. O. Chwolson, Astron. Nachr. **221,** 329 (1924).

21. A. Einstein, Astron. Nachr. **221,** 329 (1924).

22. O. Lodge, Observatory **42,** 365 (1919), and Nature **104,** 329 (1919).

23. G. C. McVittie, in *Paris Symposium on Radio Astronomy* (IAU Symp. No. 9), ed. R. N. Bracewell (Stanford, Calif.: Stanford University Press, 1959).

24. G. Gamow, Ohio J. Sci. **35,** 406 (1935); Phys. Rev. **70,** 272 (1946); Rev. Mod. Phys. **21,** 372 (1949).

25. R. A. Alpher and R. Hermann, Phys. Rev. **75,** 1092 (1949).

26. C. Hayashi, Prog. Theor. Phys. **5,** 224 (1950).

27. A. S. Eddington, *Internal Constitution of the Stars* (Cambridge: Cambridge University Press, 1926).

28. H. N. Russell, R. S. Dugan, and J. Q. Stewart, *Astronomy* (Boston: Ginn, 1926).

29. D. B. McLaughlin, *Introduction to Astronomy* (Boston: Houghton Mifflin, 1961).

30. R. H. Baker, *Introduction to Astronomy*, 6th ed. (New York: Van Nostrand, 1963)

31. S. Weinberg, *Gravitation and Cosmology* (New York: Wiley, 1973)

32. Prog. Theor. Physics, supplement no. 31 (1964).

33. V. Trimble, Publ. Astron. Soc. Pacific **96,** 1007 (1984).

34. G. Burkhardt *et al.,* eds. *Astronomy and Astrophysics Abstracts,* Vol. 50 and 49 previous volumes (Berlin: Springer-Verlag, 1989).

Man's Place in the Universe *(page 60)*

1. J. P. Ostriker, P. J. E. Peebles, and A. Yahil, Astrophys. J. **193,** 4 (1974).

2. M. G. Hauser and P. J. E. Peebles, Astrophys. J. **185,** 757 (1973).

3. W. G. Tifft and S. A. Gregory, Astrophys. J. **205,** 696 (1976).

4. E. Kellogg, S. Murray, R. Giacconi, H. Tannenbaum, and H. Gursky, Astrophys. J. **185,** L13 (1973).

5. J. R. Gott, J. E. Gunn, D. Schramm, and B. M. Tinsley, Astrophys. J. **194,** 543 (1974).

6. Ostriker *et al.,* note 1.

7. A. R. Sandage and G. Tammann, Astrophys. J. **202,** 583 (1975).

8. H. D. Curtis, J. Wash. Acad. Sci. **9,** 212 (1919); E. Hubble, Publ. Amer. Astron. Soc. **5,** 261 (1925).

9. Gott *et al.,* note 5.

10. *Ibid.*

11. P. J. E. Peebles, *Physical Cosmology* (Princeton, N. J.: Princeton University Observatory, 1971).

12. V. Trimble, Rev. Mod. Phys. **47,** 877 (1975).

13. R. Wagoner, W. A. Fowler, and F. Hoyle, Astrophys. J. **148,** 3 (1967).

14. B. Jones, Rev. Mod. Phys. **48,** 107 (1976).

15. Trimble, note 12; J. Audouze and B. M. Tinsley, Ann. Rev. Astron. Astrophys. **14,** 43–80 (1976).

16. A. S. Eddington, *The Internal Constitution of the Stars* (Cambridge: Cambridge University Press, 1926).

17. J. Bahcall and R. D. Davis, Science **191,** 264 (1976).

18. Referenced in Trimble, note 12.

19. F. Hoyle and R. A. Lyttleton, Mon. Not. R. Astron. Soc. **102,** 218 (1942), and Mon. Not. R. Astron. Soc. **109,** 614 (1949).

20. W. A. Fowler, G. Caughlan, and B. A. Zimmerman, Ann. Rev. Astron. Astrophys. **13,** 69 (1975).

21. For example, N. J. Woolf, *In Late Stages of Stellar Evolution*, IAU Symposium no. 66, Dordrecht: D. Reidel, 1974).

22. D. R. Hearn, J. A. Richardson, H. V. D. Bradt, G. W. Clark, W. H. G. Lewin, W. F. Meyer, J. E. McClintock, F. A. Primini, and S. A. Rappaport, Astrophys. J. **203,** L21 (1976).

23. W. Baade and F. Zwicky, Phys. Rev. **45,** 138 (1934), and Proc. Nat. Acad. Sci. **20,** 254 (1934).

24. D. H. Staelin and E.C. Reifenstein, Science **162,** 1481 (1968).

25. See, for example, B. Paczyński, Ann. Rev. Astron. Astrophys. **9,** 183 (1971).

26. K. S. Thorne, Sci. Am. **231** (6): 32 (1974).

27. R. P. Kirshner, and J. B. Oke, Astrophys. J. **200,** 574 (1975); M. Peimbert and S. van den Bergh, Astrophys. J. **167,** 223 (1971); M. Peimbert, Astrophys. J. **170,** 261 (1971).

28. Trimble, note 12.

29. P. van de Kamp, Ann. Rev. Astron. Astrophys. **13,** 295 (1975).

30. S. L. Miller, Science **117,** 528 (1953).

31. E. Stephen-Sherwood and J. Oró, Space Life Sci. **4,** 5 (1973).

32. E. Herbst and W. Klemperer, Phys. Today, June 1976, p. 32.

33. G. Jungclaus, J. R. Cronin, C. B. Moore, and G. V. Yuen, Nature **261,** 126 (1976), and references therein.

34. M. Calvin, Am. Sci. **63,** 169 (1975).

35. P. Cloud, Am. Sci. **62,** 54 (1974).

36. I. S. Shklovskii and C. Sagan, *Intelligent Life in the Universe* (San Francisco: Holden Day, 1966); C. Ponnamperuma and A. G. W. Cameron, *Interstellar Communication* (Boston: Houghton Mifflin, 1974); C. Sagan, *Communication with Extraterrestrial Intelligence* (Cambridge Mass: MIT Press, 1973), and references therein.

37. C. Sagan and F. D. Drake, Sci. Amer. **232** (5): 80 (1975).

38. W. Heisenberg, Phys. Today, March 1976, p. 32.

39. T. G. Van Flandern, Sci. Am. **234,** 44 (1976).

40. I. W Roxburgh, Nature **261,** 301 (1976).

41. C. B. Collins and S. W. Hawking, Astrophys. J. **180,** 317 (1973).

42. B. Carter, in *Confrontation of Cosmological Theory with Observational Data*, IAU Symposium no. 63, ed. M. S. Longair (Dordrecht: D. Reidel, 1974), p. 291.

43. S. W. Hawking and G. F. R. Ellis, *The Large Scale Structure of Space–Time* (Cambridge: Cambridge University Press, 1973).

44. Ya. B. Zel'dovich, Comm. Astrophys. Space Phys. **3,** 179 (1971).

45. C. W. Misner, K. S. Thorne, and J. A. Wheeler, *Gravitation* (New York: W. H. Freeman, 1973), ch. 44.

46. J. W. Follin, Jr., personal communication, 1976.

47. P. A. M. Dirac, Proc. Soc. **A 165,** 199 (1938); A. S. Eddington, *Fundemental Theory* (Cambridge: Cambridge University Press, 1926); M. J. Rees, Comm. Astrophys. Sp. Phys. **4,** 179 (1972); J. A. Wheeler, Am. Sci. **62,** 683 (1974).

48. R. P. Feynman, personal communication, 1975.

Classifying Ourselves *(page 115)*

1. G. de Vaucouleurs, Astrophys J. **268,** 451 and 468 (1983).

2. J. Hardorp, Astron. Astrophys. **88,** 334 (1980).

3. J. Hardorp, Astr. Astrophys. **91,** 221 (1980).

4. D. L. Barry, R. H. Cromwell, and S. A. Schoolman, Astrophys. J. **222,** 1032 (1978).

5. P. W. Hodge, Publ. Astron. Soc. Pac. **95,** 721 (1983).

6. J. Hardorp, and J. Tomkin, Astron. Astrophys. **127,** 277 (1983).

7. E. Hubble, Astrophys. J. **64,** 321 (1926).

8. A. R. Sandage, *Hubble Atlas of Galaxies* (Washington D.C.: Carnegie Institution, 1961).

9. G. de Vaucouleurs, Handb. Phys. **53,** 275 (1959).

10. W. Baade and N. U. Mayall, *1st (Paris) Symp. on Cosmical Gas Dynamics* (Daytona: Central Air Documents Office, 1949), p. 165.

11. W. Baade, Publ. Univ. Mich. Obs. **10,** 16 (1950).

12. G. de Vaucouleurs and W. Pence, Astron. J. **83,** 1163 (1983).

13. G. de Vaucouleurs, IAU Symp. **20,** 88 and 195 (1984).

14. G. de Vaucouleurs, IAU Symp. **38,** 18 (1970).

15. H. C. Arp, Astrophys. J. **139,** 1047 (1964); Arp, Astrophys. J. **141,** 43 (1965).

16. W. Becker, Z. Astrophys. **58,** 202 (1964).

17. Y. P. Georgelin and Y. M. Georgelin, Astron. Astrophys. **49,** 57 (1976).

18. G. Kuiper, Astrophys. J. **88,** 432 (1938).

19. J. Stebbins and G. E. Kron, Astrophys. J. **126,** 266 (1958).

20. L. Galloüet, C. R. Soc. Biol. **256,** 4593 (1963).

21. L. Galloüet, Ann. Astrophys. **27,** 423 (1964).

22. R. R. Emmons, *et al.,* Astron. J. **70,** 353 (1965).

23. S. van den Bergh and I. J. Sackmann, Astron. J. **70,** 353 (1965).

24. Hardorp, Astron. Astrophys. **63,** 363 (1978).

25. Hardorp, Astron. Astrophys. **96,** 123 (1981).

26. Hardorp, Astron. Astrophys. **105,** 120 (1982).

27. Y. Chmielewski, Astron. Astrophys. **93,** 334 (1981).

28. H. Tüg, and T. Schmidt-Kaler, Astron. Astrophys. **105,** 400 (1982).

29. D. Branch, D. L. Lambert, and J. Tomkin, Astrophys. J. **241,** 183 (1980).

30. V. Trimble, Rev. Mod. Phys. **55,** 511 (1983).

31. R. Armour, *The Classics Reclassified* (New York: McGraw-Hill, 1960), p. 27.

The Greatest Supernova Since Kepler *(page 167)*

1. P. Hodge and F. Wright, The Large Magellanic Cloud (V chart 54; B chart 53), 1967; P. B. Lucke and P. Hodge, Astron. J. **75,** 171 (1970).

2. N. Sanduleak, CTIO Contr. 89 (1970); J. Rousseau, N. Marin, L. Prevot, E. Rebeirot, A. Robin, and J. P. Brunet, Astron. Astrophys. Suppl. Ser. **31,** 243 (1978).

3. A. C. Fabian, M. J. Rees, E. P. J. van den Heuvel, and J. van Paradijs, Nature **328,** 323 (1987); S. R. Heap and D. J. Lindley, Astron. Astrophys. **185,** L10 (1987); P. L. Joss, P. Podsiadlowskii, J. J. L. Hsu, and S. Rappaport, Nature **331,** 237 (1988);

G. Testor, Astron. Astrophys. **190,** L1 (1988).

4. V. M. Blanco *et al.*, Astrophys. J. **320,** 589 (1987); N. R. Walborn, B. M. Lasker, V. G. Laidler, and Y.-H. Chiu, Astrophys. J. **321,** L41 (1987); R. M. West, A. Laubertz, H. E. Jurgensen, and H. E. Schuster, Astron. Astrophys. **177,** L1 (1987); G. L. White and D. F. Malin, Nature **327,** 36 (1987); T. Girard, W. F. van Altena, and C. E. Lopez, Astron. J. **95,** 58 (1988).

5. R. Gilmozzi *et al.*, Nature **328,** 318 (1987); G. Sonneborn, B. Altner, and R. P. Kirshner, Astrophys. J. **323,** L35 (1987).

6. W. D. Arnett, Astrophys. J. **319,** 136 (1987); E. K. Grasberg, V. S. Imshennik, D. K. Nadezhin, and V. P. Utrobin, Sov. Astron. Lett. **13,** 227 (1987); S. E. Woosley, P. A. Pinto, P. G. Martin, and T. A. Weaver, Astrophys. J. **318,** 664 (1987); H. Saio, M. Kato, and K. Nomoto, Astrophys. J. **331,** 388 (1988); T. Shigeyama, K. Nomoto, and M. Hashimoto, Astron. Astrophys. **196,** 141 (1988); J. C. Wheeler, R. P. Harkness, and Z. Barkat, in *Supernova 1987A in the Large Magellanic Cloud,* ed. M. Kafatos and A. C. Michalitsianos (Cambridge: Cambridge University Press, 1988), p. 264; S. E. Woosley, Astrophys. J. **330,** 218 (1988); S. E. Woosley, P. A. Pinto, and L. Ensman, Astrophys. J. **324,** 466 (1988).

7. V. Trimble, B. Paczyński, and B. Zimmerman, Astron. Astrophys. **25,** 35 (1973); W. M. Brunish and J. W. Truran, Astrophys. J. Suppl. Ser. **49,** 447 (1982); A. Maeder, Astron. Astrophys. **173,** 247 (1987).

8. Arnett, note 6; W. Hillebrandt, P. Höflich, J. W. Truran, and A. Weiss, Nature **327,** 597 (1987).

9. I. S. Shklovskii, Sov. Astron. Lett. **10,** 302 (1984).

10. B. Paczyński, Acta Astron. **20,** 195 (1970); Paczyński, Acta Astron. **22,** 163 (1971).

11. Z. Barkat, and J. C. Wheeler, Astrophys. J. **332,** 247 (1988); Saio *et al.,* note 6; Woosley, note 6.

12. A. Weiss, W. Hillebrandt, and J. W. Truran, Astron. Astrophys. Lett. **197,** L11 (1988).

13. E. Amaldi, P. Bonifazi, S. Frasca, M. Gabellieri, D. Gretz, G. V. Pallotino, G. Pizzella, J. Weber, and G. Wilmot, in *Supernova 1987A,* note 6, p. 453; O. Saavedra *et al.,* in *Proceedings of the 20th Yamada Conference, Big Bang, Active Galactic Nuclei, and Supernovae,* ed. S. Hayakawa and K. Sato (Tokyo: University Adacemy Press, 1988).

14. M. Aglietta *et al.,* Europhys. Lett. **3,** 1315 and 1321 (1987).

15. K. Hirata *et al.,* Phys. Rev. Lett. **58,** 1490 (1987).

16. R. M. Bionta *et al.,* Phys. Rev. Lett. **58,** 1494 (1987).

17. E. N. Alexeyev, L. N. Alexeyava, I. B. Krivusheina, and V. I. Polchenko, Pis'ma Zh. Eksp. Teor. Fiz. **45,** 461 [JETP Lett. **45,** 589 (1987)].

18. J. Arafune and M. Fukugita, Phys. Rev. Lett. **59,** 367 (1987); J. N. Bahcall, T. Piran, W. H. Press, and D. N. Spergel, Nature **327,** 682 (1987); N. D. Hari Dass *et al.,* Curr. Sci. **56,** 575 (1987); S. H. Kahana, J. Cooperstein, and E. Baron, Brookhaven National Laboratory Preprint No. 40012 (1987); R. Mayle, J. R. Wilson, and D. N. Schramm, Astrophys. J. **318,** 288 (1987); K. Sato and H. Suzuki, Phys. Rev. Lett.

58, 2722 (1987); Sato and Suzuki, Phys. Lett. B **196,** 267 (1987); D. N. Spergel, T. Piran, A. Loeb, J. Goodman, and J. N. Bahcall, Science **237,** 1471 (1987); Suzuki and Sato, Publ. Astron. Soc. Jpn. **39,** 521 (1987); S. Bludman and P. J. Schinder, Astrophys. J. **326,** 256 (1988).

19. D. N. Spergel and J. N. Bahcall, Phys. Lett. B **200,** 366 (1988), and references cited therein.

20. W. Hillebrandt, P. Höflich, P. Kafka, E. Müller, H. U. Schmidt, and J. W. Truran, Astron. Astrophys. **177,** L41 (1987); D. Evans, R. Fong, and P. D. B. Collins, Astron. Astrophys. **189,** 210 (1988).

21. G. B. Barbiallini and G. Cocconi, Nature **329,** 21 (1987).

22. J. M. Lattimer and M. Cooperstein, Phys. Rev. Lett. **61,** 23 (1988).

23. J. Ellis and K. A. Olive, CERN Preprint No. TH 4701/87 (1987).

24. A. Dar and S. Dado, Phys. Rev. Lett. **59,** 2768 (1987); M. Takhara and K. Sato, Mod. Phys. Lett. A **2,** 293 (1987); G. Raffelt and D. Seckel, Phys. Rev. Lett. **60,** 1793 (1988).

25. The topic is addressed in a number of preprints from CERN and Fermilab.

26. J. Arafune, M. Fukugita, T. Yanagida, and M. Yoshimura, Phys. Rev. Lett. **59,** 1865 (1987); L. M. Krauss, Nature **329,** 689 (1987).

27. M. J. Long, Phys. Rev. Lett. **60,** 127 (1988).

28. S. W. Bruenn, Phys. Rev. Lett. **59,** 938 (1987).

29. J. N. Bahcall, D. N. Spergel, and W. H. Press, in *Supernova 1987A,* note 6, p. 172.

30. A. De Rújula, Phys. Lett. B **193,** 514 (1987); W. Hillebrandt, P. Höflich, P. Kafka, E. Müller, H. U. Schmidt, and J. W. Truran, Astron. Astrophys. **180,** L20 (1987); D. N. Voskresensky *et al.,* Astrophys. Space Sci. **138,** 421 (1987).

31. E. J. Wampler, J. W. Truran, L. B. Lucy, P. Höflich, and W. Hillebrandt, Astron. Astrophys. **182,** L51 (1987).

32. T. Shigeyama, K. Nomoto, M. Hashimoto, and D. Sugimoto, Nature **328,** 320 (1987); W. D. Arnett, Astrophys. J. **331,** 337 (1988); Woosley, note 6.

33. R. Schaeffer, Y. Declais, and S. Jullian, Nature **330,** 142 (1987).

34. M. Koshiba, private communication, 1988.

35. V. Trimble, Rev. Mod. Phys. **54,** 1183 (1982).

36. Arnett, note 32; Shigeyama *et al.,* note 6; Woosley, note 6.

37. E. Baron, H. Bethe, G. E. Brown, J. Cooperstein, and S. H. Kahana, Brookhaven National Laboratory Preprint No. 39814 (1987).

38. R. H. McNaught, IAU Circular 4316 (1987); McNaught, IAU Circular 4317 (1987); McNaught, IAU Circular 4389 (1987).

39. R. M. Catchpole *et al.,* Mon. Not. R. Astron. Soc. **229,** 15P (1987); M. Hamuy, N. R. Suntzeff, R. Gonzalez, and G. Martin, Astron. J. **95,** 62 (1988).

40. P. Bouchet, R. Stanga, T. Le Bertre, N. Epchtein, W. R. Hamann, and D. Loren-
zetti, Astron. Astrophys. **177,** L9 (1987); Blanco *et al.,* note 4; S. Cristiani *et al.,*
Astron. Astrophys. **177,** L5 (1987); I. J. Danziger, R. A. E. Fosbury, D. Alloin, S.
Cristiani, J. Dachs, C. Gouiffes, B. Jarvis, and K. C. Sahu, Astron. Astrophys. **177,**
L13 (1987); R. W. Hanuschik and J. Dachs, Astron. Astrophys. **182,** L29 (1987); R.
P. Kirshner, G. Sonneborn, D. M. Crenshaw, and G. E. Nassiopoulos, Astrophys. J.
320, 620 (1987); J. W. Menzies *et al.,* Mon. Not. R. Astron. Soc. **227,** 39P (1987);
W. Wamsteker *et al.,* Astron. Astrophys. **177,** L21 (1987).

41. A. J. Turtle *et al.,* Nature **327,** 38 (1987); N. Bartel *et al.,* in *Supernova 1987A,* note
6, p. 81.

42. R. A. Chevalier and C. Fransson, Nature **328,** 44 (1987); M. C. Storey and R. N.
Manchester, Nature **329,** 421 (1987).

43. Hillebrandt *et al.,* note 8; R. Schaeffer, M. Casse, R. Mochkovitch, and S. Cahew,
Astron. Astrophys. **184,** L1 (1987); Shigeyama *et al.,* note 32; Arnett, note 32; Arnett, in *Supernova 1987A,* note 6, p. 301; Woosley, note 6; Woosley, in *Supernovae
1987A,* note 6.

44. D. Branch, Astrophys. J. **320,** 421 (1987); L. B. Lucy, Astron. Astrophys. **182,** L31
(1987); P. Höflich, 1988 (forthcoming in *Atmospheric Diagnostics of Stellar Evolution,* IAU Colloquium No. 108, ed. K. Nomoto, Lecture Series in Physics [Berlin:
Springer-Verlag]); L. B. Lucy, in *Supernova 1987A,* note 6, p. 323; Wheeler *et al.,*
note 6.

45. A. C. Raga, Astron. J. **94,** 1578 (1987).

46. M. Dopita, S. J. Meatheringham, P. Nulsen, and P. R. Wood, Astrophys. J. **322,**
L85 (1987); Woosley, note 6.

47. Höflich, note 44; Lucy, note 44; Wheeler *et al.,* note 6.

48. B. N. Ashoka, G. C. Anupama, T. P. Prabhu, S. Giridhar, K. K. Ghosh, S. K. Jain,
A. K. Pati, and N. Kameswara Rao, J. Astrophys. Astron. **8,** 195 (1987); E. Oliva,
A. F. M. Moorwood, and I. J. Danziger, ESO Messenger **50,** 18 (1988); J. A. Tyson
and P. Boeshaar, Publ. Astron. Soc. Pac. **99,** 905 (1987); R. E. Williams, Astro-
phys. J. **320,** L117 (1987); D. K. Aitken, C. H. Smith, S. D. James, P. I. Roche, H.
R. Hyland, and P. J. McGregor, Mon. Not. R. Astron. Soc. **231,** 7P (1988); R. M.
Catchpole *et al.,* Mon. Not. R. Astron. Soc. **231,** 75P (1988); P. Harvey *et al.,* IAU
Circular 4518 (1988); M. M. Phillips, Astron. J. **95,** 1087 (1988); Phillips, in *Supernova 1987A,* note 6, p. 16; P. A. Whitelock *et al.,* Mon. Not. R. Astron. **234,** 5p
(1988).

49. Arnett, note 32; Woosley, note 6.

50. Woosley, note 6.

51. T. A. Weaver and S. E. Woosley, Ann. N.Y. Acad. Sci. **336,** 335 (1980).

52. A. Uomoto, and R. P. Kirshner, Astrophys, J. **308,** 685 (1986).

53. Phillips, note 48, both works.

54. L. B. Lucy, in *Supernova 1987A,* note 6, p. 323.

55. J. E. Gunn and J. P. Ostriker, Nature **221,** 454 (1968).

56. R. McCray, J. M. Shull, and P. Sutherland, Astrophys. J. **317,** L73 (1987); J. P. Ostriker, Nature **327,** 287 (1987); Shigeyama *et al.,* note 32.

57. Williams, note 48.

58. Höflich, note 44.

59. Catchpole *et al.,* note 48.

60. Danziger *et al.,* note 40.

61. Catchpole *et al.,* note 48.

62. Aiken *et al.,* note 48; Harvey *et al.,* note 48; Oliva *et al.,* note 48.

63. H. Moseley, E. Dwek, W. Glaccum, J. Graham, R. Loewenstein, and R. Silverberg, IAU Circular 4576 (1988).

64. W. Wamsteker, R. Gilmozzi, A. Cassatella, and N. Panagia, IAU Circular 4410 (1987); R. P. Kirshner, in *Supernova 1987A,* note 6, p. 87.

65. C. Fransson, A. Cassetella, R. Gilmozzi, R. Kirshner, N. Panagia, G. Sonneborn, and W. Wamsteker, 1988 (forthcoming in Astrophys. J.).

66. R. A. Chevalier, Nature **332,** 514 (1988).

67. Frannson *et al.,* note 65.

68. M. Itoh, S. Hayakawa, K. Masai, and K. Nomoto, Publ. Astron. Soc. Jpn. **39,** 529 (1987).

69. Whitelock *et al.,* note 48.

70. M. Hamuy and M. Phillips, IAU Circular 4534 (1988).

71. C. Fransson and R. A. Chevalier, 1988 (forthcoming in Astrophys. J.).

72. P. A. Pinto, S. E. Woosley, and L. M. Ensman, Astrophys. J. **331,** L101 (1988).

73. F. C. Michel, C. F. Kennel, and W. A. Fowler, Science **238,** 938 (1987); R. Bandiera, F. Pacini, and M. Salvati, Nature **332,** 418 (1988); Fransson and Chevalier, note 71.

74. K. Nomoto and S. Tsuruta, in *Supernova 1987A,* note 6, p. 421.

75. K. W. Chan and R. E. Lingenfelter, Astrophys. J. **318,** L51 (1987); N. Gehrels, E. J. MacCallum, and M. Leventhal, Astrophys. J. **320,** L19 (1987).

76. S. A. Grebenev and R. A. Sunyaev, Pis'ma Astron. Zh. **13,** 945 (1987) [Sov. Astron. Lett. **13,** 397 (1987)]; Grebenev and Sunyaev, Pis'ma Astron. Zh. **13,** 1042 (1987) [Sov. Astron. Lett. **13,** 438 (1987)]; McCray *et al.,* note 56; Y. Xu, P. Sutherland, R. McCray, and R. R. Ross, Astrophys. J. **327,** 197 (1988).

77. T. Dotani *et al.,* Nature **330,** 230 (1987); F. Makino and the Ginga Team, IAU Circular 4447 (1987); R. A. Sunyaev *et al.,* Nature **330,** 227 (1987).

78. M. Itoh, S. Kunagai, T. Shigeyama, K. Nomoto, and J. Nishimura, Nature **330,** 233 (1987); T. Ebisuzaki and N. Shibazaki, Astrophys. J. **327,** L5 (1988); Ebisuzaki and Shibazaki, Astrophys. J. **328,** 699 (1988).

79. Woosley, note 6.

80. Arnett, note 43; Woosley, note 43.

81. Shibazaki and Ebisuzaki, note 78, **327.**

82. S. M. Matz *et al.*, IAU Circular 4510; Matz *et al.*, IAU Circular 4568 (1988).

83. W. R. Cook, D. Palmer, T. Prince, C. Schindler, C. Starr, and E. Stone, IAU Circular 4527 (1987).

84. W. Sandie, G. Nakano, L. Chase, G. Fishman, C. Meegan, R. Wilson, W. Paciesas, and G. Lasche, IAU Circular 4526 (1988).

85. A. C. Rester, G. Eichborn, R. L. Coldwell, J. I. Trombka, R. Starr, and G. P. Lasche, IAU Circular 4535 (1988).

86. W. D. Arnett and A. Fu, 1988 (forthcoming in Astrophys. J.); A. Fu and W. D. Arnett, 1988 (forthcoming in Astrophys. J.); P. A. Pinto and S. E. Woosley, Nature **333,** 534 (1988).

87. Sandie *et al.,* note 84.

88. R. E. Lingenfelter, private communication, 1988.

89. M. D. Leising, Nature **332,** 516 (1988).

90. Rester *et al.,* note 8.

91. Pinto and Woosley, note 86.

92. F. Makino and the Ginga Team, IAU Circular 4530 (1988); Makino and Ginga Team, IAU Circular 4532 (1988).

93. R. A. Sunyaev *et al.,* preprint (1988) in Russian; P. Ubertini *et al.,* IAU Circular 4590 (1988).

94. Y. Tanaka, in *Proceedings of the 20th Yamada Conference,* note 13.

95. P. Bouchet and I. J. Danziger, IAU Circular 4575 (1988).

96. H. Moseley, W. Glaccum, R. Loewenstein, R. Silverberg, E. Dwek, and J. Graham IAU Circular 4500 (1987); Moseley *et al.,* note 63; M. R. Haas, S. W. J. Colgan, E. F. Erickson, S. D. Lord, and M. G. Burton, IAU Circular 4578 (1988); H. Moseley, E. Dwek, W. Glaccum, J. Graham, R. Loewenstein, and R. Silverberg, IAU Circular 4574 (1988); D. M. Rank, P. A. Pinto, S. E. Woosley, J. D. Bregman, F. C. Witteborn, T. Axelrod, and M. Cohen, Nature **331,** 505 (1988); F. Witteborn, J. Bregman, D. Wooden, P. Pinto, D. Rank, and M. Cohen, IAU Circular 4592 (1988).

97. W. H. Couch, in *Supernova 1987A,* note 6, p. 60.

98. Bandiera *et al.,* note 73.

99. H. E. Schwarz and R. Mundt, Astron. Astrophys. **177,** L4 (1987); P. Barrett, SAAO Preprint (1988); Couch, note 97; M. W. Feast, in *Supernova 1987A,* note 6, p. 51.

100. D. T. Jeffery, Nature **329,** 419 (1987).

101. Couch, note 97.

102. M. Cropper, J. Bailey, J. McCowage, R. D. Cannon, W. J. Couch, J. R. Walsh, J. O. Straede, and F. Freeman, Mon. Not. R. Astron. Soc. **231,** 695 (1988).

103. Barrett, note 99.

104. Ashoka *et al.,* note 48; Hanuschik and Dachs, note 40; R. W. Hanuschik, G. Theim, and J. Dachs, 1988 (forthcoming in Mon. Not. R. Astron. Soc.); H. P. Larson, in *Supernova 1987A,* note 6, p. 74; Phillips, note 48, both papers.

105. L. B. Lucy, in *Supernova 1987A* note 6, p. 323.

106. Haas *et al.*, note 96; Rester *et al.*, note 85; Witteborn *et al.*, note 96.

107. Hanuschik *et al.*, note 104.

108. Phillips, note 48, both papers.

109. W. P. S. Meikle, S. S. Matcher, and B. L. Morgan, Nature **329**, 608 (1987); P. Nisenson, C. Papaliolios, M. Karovska, and R. Noyes, Astron. J. **320**, L15 (1987).

110. S. Matcher *et al.*, IAU Circular 4543 (1988); C. Papaliolios *et al.*, in *Supernova 1987A*, note 6, p. 225.

111. M. J. Rees, Nature **328**, 207 (1987).

112. I. Goldman, Astron. Astrophys. **186**, L3 (1987).

113. W. Hillebrandt, P. Höflich, H. U. Schmidt, and J. W. Truran, Astron. Astrophys. **186**, L9 (1987); E. S. Phinney, Nature **329**, 698 (1987).

114. M. Karovska, L. Koechlin, P. Nisenson, C. Papaliolios, and C. Standley, IAU Circular 4521 (1987); Karovska, Koechlin, Nisenson, Papaliolios, and Standley, IAU Circular 4604 (1988); Papaliolios *et al.*, note 110.

115. A. Chalabaev, C. Perrier, and J. M. Mariotti, IAU Circular 4481 (1987).

116. E. Dwek, in *Supernova 1987A*, note 6, p. 240, and Astrophys. J. **329**, 814 (1988).

117. A. Crotts, IAU Circular 4561 (1988); C. Gouiffes, M. Rosa, J. Melnick, I. J. Danziger, M. Remy, C. Santini, J. L. Sauvageot, P. Jakobsen, and M. T. Ruiz, Astron. Astrophys. **198**, L9 (1988); S. Heathcote and N. R. Suntzeff, IAU Circular 4567 (1988); M. Rosa, IAU Circular 4564 (1988).

118. Heathcote and Suntzeff, note 117.

119. Crotts, note 117.

120. P. Coudere, Ann. Astrophys. **2**, 271 (1939).

121. B. E. Schaefer, Astrophys. J. **323**, L47 (1987).

122. C. R. Canizares, G. A. Kriss, and E. D. Feigelson, Astrophys. J. **253**, L17 (1982).

123. K. W. Weiler, R. A. Sramek, J. M. van der Hulst, and N. Panagia, in *Supernovae: A Survey of Current Research*, ed. M. J. Rees and R. H. Stoneham (Dordrecht: Reidel, 1982), p. 281.

124. R. A. Chevalier, Astrophys. J. **259**, 302 (1982).

125. Makino *et al.*, note 77, and note 92, both papers; Tanaka, note 94; Tanaka, in *Supernova 1987A*, note 6, p. 349.

126. Tanaka, in *Supernova 1987A*, note 6, p. 349.

127. B. Aschenbach, Nature **330**, 232 (1987); U. G. Briel, E. Pfeffermann, B. Aschenbach, H. Bräuninger, and J. Trümper, IAU Circular 4452 (1987); D. Burrows, J. Nousek, and G. Garmire, IAU Circular 4494 (1987).

128. Tanaka, in *Supernova 1987A*, note 6, p. 349.

129. Hillebrandt *et al.*, note 113.

130. K. Masai, S. Hayakawa, H. Itoh, and K. Nomoto, Nature **330**, 235 (1987); Masai, in *Proceedings of the 20th Yamada Conference*, note 13.

131. Masai, *Proceedings of the 20th Yamada Conference,* note 13.

132. Bandiera *et al.,* note 73.

133. I. A. Bond *et al.,* Phys. Rev. Lett. **60,** 1110 (1988); D. Ciampa *et al.,* Astrophys. J. **326,** L9 (1988).

134. V. S. Berezinsky and V. I. Ginzberg, Nature **329,** 807 (1987); M. Honda and M. Mori, Prog. Theor. Phys. **78,** 63 (1987); Honda and Mori, Prog. Theor. Phys. **78,** 1065 (1987); R. J. Protheroe, Nature **329,** 135 (1987).

135. T. Kifune, in *Proceedings of the 20th Yamada Conference,* note 13.

136. G. Srinivasan, in *Supernovae, Their Progenitors and Remnants*, ed. G. Srinivasan and V. Radhakrishnan (Bangalore: Indian Academy of Science, 1985), p. 105.

137. Nomoto and Tsuruta, note 74.

So You Want to Find a Supernova? *(page 187)*

1. V. Trimble, Rev. Mod. Phys. **54,** 1183 (1982); Trimble, Rev. Mod. Phys. **55,** 511 (1983); A. Petschek, ed., *Supernovae* (Berlin: Springer-Verlag, forthcoming); S. E. Woosley, ed., *Supernovae* (Berlin: Springer-Verlag, forthcoming); J. C. Wheeler *et al.,* eds., *Supernovae* (World Scientific, forthcoming).

2. R. Barbon, E. Cappellaro, and M. Turatti, Astron. Astrophys. Sup. **81,** 421 (1989).

3. *Ibid.*

4. S. Bruenn, Astrophys. J. **341,** 489 (1989); E. S. Myra and S. Bludman, Astrophys. J. **340,** 384 (1989).

5. S. A. Colgate, Nature **341,** 489 (1990).

6. J. C. Wheeler and R. P. Harkness, 1990 (forthcoming in Rep. Prog. Phys.).

7. P. Bergeron *et al.,* Astrophys. J. **345,** L91 (1989).

8. E. M. Schlegel and R. P. Kirshner, Astron. J. **98,** 577 (1989); R. A. Chevalier, Astrophys. J. **343,** 323 (1989).

9. Wheeler and Harkness, note 6.

10. O.-G. Richter and M. Rosa, Astron. Astrophys. **206,** 219 (1989).

11. S. van den Bergh, Astron. Astrophys. **231,** L27 (1990).

12. L. C. O'Drury, Astron. Astrophys. **225,** 175 (1989).

13. G. Tenorio-Tagle *et al.,* eds., *Structure and Dynamics of the ISM* (Berlin: Springer-Verlag, 1989).

14. F. Matteuchi and A. Tornambé, Comm. Astrophys. **12,** 245 (1988).

15. R. P. Kirshner, Nature **339,** 512 (1989); H. U. Nørgaard-Nielson *et al.,* Nature **339,** 523 (1989); D. H. Hauscholdt *et al.,* Astron. Astrophys. **210,** 262 (1989); W. Schmutz *et al.,* Astrophys. J. **355,** 255 (1990); B. Leibundgut and G. A. Tammann, Astron. Astrophys. **230,** 81 (1990).

16. B. Leibundgut, Astron. Astrophys. **229,** 1 (1990).

17. H. C. Arp, Publ. Astron. Soc. Pac. **102**, 436 (1990).

18. E. K. Grasberg and D. K. Nadezhin, Sov. Astron. **31**, 629 (1987).

19. G. M. Voit, Astrophys. J. **331**, 343 (1988).

20. M. Takahara and K. Sato, Astrophys. J. **335**, 301 (1989).

21. I. S. Shklovskii, Sov. Astron. Lett. **10**, 302 (1984).

22. W. Ashworth, J. Hist. Astron. **11**, 1 (1980).

23. G. A. Tammann, in M. J. Rees and R. Stoneham, eds., *Supernovae* (Dordrecht: Reidel, 1982), p. 371.

24. E. Cappellaro and M. Turatti, Astron. Astrophys. **190**, 1 (1988); R. Evans, S. van den Bergh, and R. D. McClure, Astrophys. J. **345**, 752 (1989).

25. K. Ratnatunga and S. van den Bergh, Astrophys. J. **343**, 713 (1989).

26. D. O. Wood and E. Churchwell, Astrophys. J. **340**, 265 (1989).

27. S. van den Bergh, Astron. J. **99**, 843 (1990).

28. V. Trimble and D. H. Clark, Bull. Astron. Soc. India **13**, 117 (1985).

29. S. T. Dye *et al.*, Phys. Rev. Lett. **62**, 2069 (1989).

30. S. van den Bergh and R. D. McClure, Astrophys. J. **347**, L29 (1989).

31. R. Mayle and J. R. Wilson, Astrophys. J. **334**, 906 (1988).

32. D. L. Miller and D. Branch, 1990 (preprint).

33. S. Newcomb and E. S. Holden, *Astronomy* (Boston: Ginn, 1879), p. 244

Remnants of Historical Supernovae *(page 195)*

1. D. H. Clark and F. R. Stephenson, *The Historical Supernovae* (New York: Pergamon, 1977).

2. See the discussion by K. Brecher, in *The Crab Nebula and Related Supernova Remnants,* ed. M. C. Kafatos and R. C. B. Henry (Cambridge: Cambridge Univiversity. Press, 1985); Clark and Stephenson, note 1, and references within.

3. J. Maza and S. van den Bergh, Astrophys. J. **204**, 519 (1976).

4. K. W. Kamper and S. van den Bergh, IAU Symp. No. 101, p. 55 (1983).

5. W. B. Ashworth, J. Hist. Astron. **11**, 1 (1980).

6. G. Tammann, in *Supernovae: A Survey of Current Research,* ed. M. J. Rees and R. J. Stoneham (Dordrecht: Reidel, 1982), p. 371; Clark and Stephenson, note 1.

7. D. H. Clark, in *Supernovae, Their Progenitors and Remnants,* ed. V. Radhakrishnan and G. Srinivasan (Bangalore: Indian Academy of Sciences, 1985).

8. R. Minkowski, Publ. Astron. Soc. Pac. **82**, 383, 470, 480 (1970); V. Trimble, Publ. Astron. Soc. Pac. **82**, 375, 480 (1970); L. Woltjer, Publ. Astron. Soc. Pac. **82**, 479 (1970).

9. L. Rosenfeld, *Astrophysics and Gravitation: Proc. 16th Solvay Conf.*, University of Bruxelles (1983), p. 174.

10. W. Glaccum, D. A. Harper, R. F. Loewenstein, R. Pernic, and F. J. Low, Bull. Am. Astron. Soc. **14**, 612 (1982).

11. K. W. Weiler, in *The Crab Nebula*, note 2; Weiler, in *Supernovae*, note 7.

12. D. J. Helfand and R. H. Becker, Nature **307**, 215 (1983).

13. G. Srinivasan, D. Bhattacharya, and K. S. Dwarakanath, J. Astrophys. Astron. **5**, 403 (1984).

14. M. C. Kafatos and R. C. B. Henry, eds., *The Crab Nebula*, note 2.

15. S. van den Bergh, Astrophys. J. Lett. **160**, L27 (1970).

16. R. A. Fesen and T. R. Gull, in *The Crab Nebula*, note 2.

17. P. Shull, U. Carsenty, M. Sarcander, and T. Neckel, in *The Crab Nebula*, note 2.

18. T. Velusamy, in *The Crab Nebula*, note 2.

19. R. C. B. Henry, in *The Crab Nebula*, note 2.

20. D. H. Clark, P. Murdin, R. Wood, R. Gilmuzzi, J. Dangizer, and A. M. Furr, Mon. Not. R. Astron. Soc. **204**, 415 (1983).

21. D. H. Clarke and F. R. Stephenson, in M. J. Rees and R. J. Stoneham, eds., *Supernovae: A Survey of Current Research*, p. 355 (1982).

Gravitational Radiation and the Binary Pulsar *(page 207)*

1. J. H. Taylor and J. M. Weisberg, Astrophys. J. **253**, 908 (1982).

2. J. N. Bahcall, Ann. Rev. Astron. Astrophys. **16**, 241 (1978).

3. R. P. Kraft, J. Mathews, and J. L. Greenstein, Astrophys. J. **136**, 312 (1962); V. B. Braginski, Usp. Fiz. Nauk **86**, 433 (1965); B. Paczyński, Acta Astron. **17**, 207 (1967); J. Faulkner, Astrophys. J. Lett. **9**, 183 (1971); J. Pringle and R. F. Webbink, Mon. Not. R. Astron. Soc. **172**, 493 (1975).

4. B. Paczyński, Ann. Rev. Astron. Astrophys. **9**, 183 (1971).

5. B. Paczyński and R. Sienkiewicz, Astrophys. J. **248**, L27 (1981); Paczyński and Sienkiewicz, preprint (Orange Aid Series, 1982); S. Rappaport, P. C. Joss, and R. F. Webbink, Astrophys. J. **254**, 616 (1982).

6. A. Einstein, Sber. preuss. Akad. Wiss. Phys.-Math. **K1**, 154 (1918).

7. L. Landau and E. Lifschitz, *Classical Theory of Fields* (Reading, Mass.: Addison-Wesley, 1951); M. Walker and C. M. Will, Astrophys. J. Lett. **242**, L129 (1980), and references therein; J. M. Cohen, Ann. N.Y. Acad. Sci. **375**, 459 (1982), and references therein.

8. Cohen, note 7; J. Ehlers, A. Rosenblum, J. N. Goldberg, and P. Havas, Astrophys. J. Lett. **208**, L77 (1976); F. I. Cooperstock and D. W. Hobill, Astrophys. J. Lett. **235**, L55 (1980).

9. J. M. Weisberg and J. H. Taylor, Gen. Relativ. Gravit. **13**, 1 (1981).

10. V. Ferrari, G. Pizzella, M. Lee, and J. Weber, Phys. Rev. **D25**, 2471 (1982).

Masada, Suicide, and Halakhah *(page 210)*

1. Marvin David Levy, *Masada* (New York: Boosey & Hawks, 1973).

2. This is the 1917 Jewish Publication Society version. Others vary almost beyond recognition. For instance, D. Novak in chapter 9 of *Law and Theology in Judaism* (New York: Ktav Publications, 1974) renders the opening phrase as "For your lifeblood too I will require a reckoning."

3. Not to be confused with some dozens of other Rabbis Eleazar (Eliezar, Elazar) who appear in various relevant talmudic passages and elsewhere here.

4. *Baba Kama* 91b. All talmudic passages cited here are from the translation *Babylonian Talmud*, edited by I. Epstein (London: Soncino Press, 1935–48).

5. Cited, e.g., by F. Rosner, *Tradition 11*, no. 2, p. 25 (1970).

6. The JPS translation reads "for that he sinned by reason of the dead" and appears to refer to a Nazarite who has accidentally been defiled because a man died suddenly in his vicinity. This would not readily admit of Rabbi Eleazar Hakkapar's interpretation.

7. According to H. H. Graetz, *History of the Jews* (Jewish Publication Society, 1891–98, reprinted 1967), vol. II, pp. 423–24, or a few years earlier, according to I. Halevy, *Dorot ha-Rishonim* (Pressburg and Frankfurt, 1897–1913), p. 371.

8. These events have been discussed from a legal point of view by Rosner, note 5; D. I. Frimer, *Tradition 12*, no. 1, p. 17 (1971); L. I. Rabinowitz, *Tradition 11*, no. 3, p. 31 (1970); S. B. Hoenig, *Tradition 11*, no. 1, p. 5 (1970), and *Tradition 13*, no. 1, p. 100 (1972); S. Spero, *Tradition 11*, no. 1, p. 31 (1970); and A. Kolitz, *Tradition 12*, no. 1, p. 5 (1970). Many of the references to the later Rabbinic literature are taken from these discussions.

9. Translation by D. Zlotnick, *The Tractate Mourning (Semahot)* (New Haven: Yale University Press, 1966). This name ("joys") for the tractate (originally known as *Evel Rabbati*—great tractate on mourning) seems akin to the Greek name for the Furies (Eumenides) and somehow un-Jewish.

10. On the basis of the same incidents, the Sages also conclude that an erring child should be punished or forgiven at once but not threatened. This is, of course, sound practice in training any living creature.

11. Novak, note 2.

12. *Mishnah Torah*, Laws of Mourning, chap. 1, sec. 11.

13. *Shulhan Arukh*, sec. *Yoreh de'ah* 345.

14. *Tur*, sec. *Yoreh de'ah* 345.

15. Nahmanides, "Torat Ha Adam," *Kitbei Ramban* 2, pp. 83–84.

16. E.g., a note of Rabbi Shabati Ha-Kohen (Shak) to *Yoreh de'ah* 345.1, referring to Rabbi Solomon ibn Adret (Rashba), *Responsa,* no. 763; Eisenstadt, *Pit'hei Teshubah* to *Y.D.* 345.1

17. Sanhedrin 74a.

18. *Genesis Rabbah* 34:12, as translated by H. Freedman, *Midrash Rabbah-Genesis* (London: Soncino Press, 1939).

19. *Ibid.*

20. Novak, note 2.

21. The story is told in Daniel 3:14-25.

22. Sanhedrin 74a.

23. According, of course, to Rabbi Eliezar in Sanhedrin 74a.

24. Sanhedrin 74a.

25. Sanhedrin 74b.

26. The same passage also defines some cases where, despite first impressions, there is no serious transgression, thus no responsibility to suffer death. For instance, a Jew is permitted to kindle a fire in a heathen temple, provided that the purpose of the fire is not worship of the false god but only to keep the heathens warm. And Esther's sin in accepting the uncircumcised Ahasuerus as her husband was not great, as she was only "natural soil (which is tilled)," *i.e.,* the passive object of his embraces. (But presumably she was forbidden to enjoy it.)

27. Of whom the best known were Akiba and Gamliel, but of course there was also an Eleazar (ben Shammua).

28. Some of the statistics of what happened to Jewish Council members are discussed in I. Trunk, *Judenrat* (New York: Macmillan, 1972).

29. Nahmanides, note 15.

30. Radak on I Samuel 31:4; Nahmanides, note 15.

31. Abravanel on I Samuel 31:7, *Knesset Ha-Gedolah* on *Yoreh de'ah* 157; *Yephei Toar* on *Genesis Rabbah* 34:12.

32. Maharshal on *Gittin* 57b.

33. Maharshal, *Yam Shel Shlomo, Baba Kama* 8:59; Semak, *Sefer Amudei Golah; Perush ha-Maharzav* on *Genesis Rabbah* 34:12.

34. See, e.g., Frimer, note 8: his note 16b.

35. *Or Ha-Mizrach, Tamaz-Elul* 5720.

36. E.g., Abravanel on I Samuel 31:4; Rabbi David Kimhi (Radak) on I Samuel 31:4; the codes as per notes 12, 13, and 14; *Genesis Rabbah* 34:15; and many others cited by Frimer, note 8: his notes 9, 11, 34, 35, 36.

37. Ritva on the *Eyn Ya'akov Avodah Zarah* 18a.

38. Ritva, note 37; *Tosafot Rash M'Shant Avodah Zarah* 18a; and many others cited by Frimer, note 8: his note 26.

39. Quoted in *Tosafot Avodah Zarah* 18:19.

40. Jewish Wars, Book 7. Yossipon is clearly not independent, and while his version of events could, by chance, have been more nearly correct, it is not the one that most writers on Masada seem to have in mind.

41. See Hoenig, Spero, and Kolitz, in note 8, and references therein.

42. Josephus, Antiquities, 18.1.

43. For instance, the discussion of *Pesahim* 25b and *Tosephta Terumot* 7:23 in L. Jacobs, *Jewish Law* (New York: Behrman House, 1968), pp. 79–83.

44. See also rulings of Rabbi I. Yehuda Unterman (*Shevet M'Yehuda*, gate 1, chap. 12, no. 1) and Rabbi Yaakov Emden (basing himself on *Radvaz*), cited by Frimer, note 8, his notes 49 and 50.

45. *Perahat Derakhim "Kerekh Ha-Rabbim,"* Sermon 13.

46. Abravanel on II Samuel 1:14.

47. *Minhat Hinukh*, Commandment 34, quoting Rambam, chap. 2 of The Laws of Murder.

48. For instance, Rabbi Meir of Rothenburg (*She'elot U'Teshuvot Maharam M'Ruttenberg*, vol. I, no. 59, ed. Moshe Aryeh Block) and the Ralbag, *Minhat Hinukh*, and Abravanel on II Samuel 1:14.

49. For instance, *Da'at Z'keinim M'ba'alei Ha-Tosafot, Orchot Ha-Haim*, and *Bet Joseph* on II Samuel 1:14, and Rabbi Meir, note 48.

50. What did you expect?

51. *Or Ha-Mizrach,* Tamuz-Elul 5720.

52. As marshaled, e.g., by Frimer, note 8.

53. *Arukh Ha-Shulhan, Yoreh de'ah* 345:5.

54. Y. Yadin, *Masada: Herod's Fortress and the Zealot's Last Stand* (New York: Random House, 1966); Spero, note 8.

55. Rabinowitz, note 8.

The Best Is Yet to Be *(page 235)*

1. H. A. Abt, Publ. Astron. Soc. Pac. **94**, 213 (1982); H. A. Abt, Publ. Astron. Soc. Pac. **95**, 113 (1983).

2. Abt, note 1, 1983; Abt, Publ. Astron. Soc. Pac. **93**, 207 (1981).

3. I am indebted to H. A. Abt for additional information about the papers and individuals discussed in notes 1 and 2.

Information Explosion *(page 238)*

1. H. A. Abt, Publ. Astron. Soc. Pac. **93**, 269 (1981).
2. W. E. Harris, Publ. Astron. Soc. Pac. **95**, 989 (1983).
3. D. L. Rector, private communication.

Progress Is Not Our Most Important Product *(page 242)*

1. M. Struble, and C. Ftaclas, Publ. Astron. Soc. Pac. **94**, 763 (1982).

Death Comes as the End *(page 244)*

1. A. R. Rao and M. N. Vahia, Publ. Astron. Soc. Pac. **98**, 511 (1986).
2. H. A. Abt, Publ. Astron. Soc. Pac. **96**, 746 (1984).
3. H. A. Abt, Publ. Astron. Soc. Pac. **95**, 113 (1983).
4. L. Woltjer, personal communication (1984).
5. H. A. Abt, Publ. Astron. Soc. Pac. **93**, 207 (1981).
6. V. Trimble, Quart. J. R. Astron. Soc. **26**, 40 (1985).
7. *Ibid.*
8. Rao and Vahia, note 1.

Young Versus Established American Astronomers *(page 250)*

1. Q. H. Flaccus (Horace), *Odes,* Book III, no. 6, lines 46–48 (*c.* 25 B.C.E.).
2. V. Trimble, Quart. J. R. Astron. Soc. **26**, 40 (1985).
3. *Ibid.*

Prestigious Start *(page 261)*

1. J. Gourman, *The Gourman Report* (Los Angeles: National Education Standards, 1987).
2. U. Esser *et. al.,* eds., *Astronomy and Astrophysics Abstracts,* Vol. 45 and earlier volumes (Heidelberg: Springer-Verlag, 1989).

Self-Citation Rates in Astronomical Papers *(page 265)*

1. A. R. Rao and M. N. Vahia, Publ. Astron. Soc. Pac. **98,** 511 (1986).
2. H. A. Abt, Publ. Astron. Soc. Pac. **92,** 249 (1980); Abt, personal communication (1986).

Acknowledgements

CHEOPS' PYRAMID was originally published as "Astronomical Investigation Concerning the So-Called Air-Shafts of Cheops' Pyramid" in *Mitteilungen des Instituts für Orientforschung der Deutschen Akademie der Wissenschaften zu Berlin,* volume 10 (1964),pages 183-87.

STAR OF BETHLEHEM was originally published as a book review of *The Star of Bethlehem: An Astronomer's Confirmation* by David Hughes, in *Archaeoastronomy,* volume 3 (1980), No. 3, page 26.

IN THEIR OWN WORDS originally appeared in *After the First Three Minutes,* the proceedings of a conference edited by S. Holt, C. Bennett, and V. Trimble and published by the American Institute of Physics in 1991. The talk was, in principle, based entirely on published papers, but the authors's thinking has been heavily influenced by seminars, private conversations, and other modes of communication. In particular, she wishes to thank for their various insights George O. Abell, Horace W. Babcock, George Gamow, James E. Gunn, Vera C. Rubin, Joseph Weber, Olin C. Wilson, Yakov B. Zeldovich, and Fritz Zwicky.

RELATIVISTIC ASTROPHYSICS had its origins in a talk that organizers Ken Brecher and Clifford Will asked the author to give at a session of the Washington meeting of the American Physical Society marking the 75th anniversary of general relativity. The full paper, under the title "Some Notes on Relativistic Astrophysics: Its History, Literature, and Demographics," was prepared in 1991 for a festschrift honoring I. N. Sengupta and edited by A. V. Wyatt for publication in 1992.

ACCORDING TO HOYLE reports a symposium held July 15-19, 1974 in Venice to celebrate the 60th birthday of Sir Fred Hoyle. It was first published as "Frontiers of Astronomy" in *Science,* volume 190 (1975), pages 368-9.

ARE GALAXIES HERE TO STAY?, written with Martin J. Rees, originally appeared in *Astronomy,* July 1978, pages 53-59.

MAN'S PLACE IN THE UNIVERSE was originally published as "Cosmology: Man's Place in the Universe" in *American Scientist,* volume 65 (1987),pages 76-85. The author is grateful to the Aspen Center for Physics, where much of this piece was written, for hospitality, and to the Alfred P. Sloan Foundation for a Research Fellowship.

The author's thoughts on the Universe and its contents have inevitably been influenced by more people than can conveniently be mentioned, but those from whom she first heard some of the ideas discussed here, and whom she would therefore like to thank especially, include Dave Arnett, William A. Fowler, Jim Gunn, Philip Morrison, Paula Moddel, Bohdan Paczyński, Martin Rees, Bill Saslaw, Starling Trimble, and (last only in this deliberately alphabetical list), Joe Weber.

DARK MATTER is the original version of a paper that was edited and published in the August 1987 issue of *Astronomy* as "The Search for Dark Matter." Reproduced by permission of publisher. Copyright 1987 by Kalmbach Publishing Company.

THE ANTHROPIC PRINCIPLE was commissioned by *The World and I* in May 1987 under the title "The Anthropic Principle as a Scientific Tool" but never published. It appears here for the first time.

IT'S A NICE PLANET TO VISIT, BUT I WOULDN'T WANT TO LIVE THERE originally appeared in *Cosmic Search,* Winter 1980, pages 13-16. Its motivation was a request from the editor for an article at a time when the author was teaching geophysics for the first time and so thinking about plate tectonics and its significance.

WHERE ARE THEY? is a report on a Symposium on the Implications of Our Failure to Observe Extraterrestrials held on November 2 and 3, 1979, at the University of Maryland. It originally appeared in *Cosmic Search,* Spring 1980, pages 20-24. The author participated in the symposium as a last-minute replacement for Beatrice M. Tinsley, who generously contributed material for the published proceedings.

CLASSIFYING OUR STAR AND GALAXY was originally published as "Classifying ourselves" in *Nature,* volume 308 (1984), pages 407-408. Reprinted by permission. Copyright 1984 by Macmillan Magazines Ltd.

THE ODD TWO PERCENT was originally published as "The Odd Two Percent: Chemical Evolution of Galaxies" in *Astronomy Now,* January 1990, pages 28-32. The idea for such an article came from Kevin Marshall, then of Medellin, Columbia.

CLOSE BINARY STARS was originally published as "A Field Guide to Close Binary Stars" in *Sky & Telescope,* volume 68 (1984), pages 306-311.

CATACLYSMIC VARIABLES was originally published as "How to Survive the Cataclysmic Binaries" in *Mercury,* January-February 1980, pages 8-12.

WHITE DWARFS: THE ONCE AND FUTURE SUNS was published in an edited version by *Sky & Telescope,* volume 72 (1986), pages 348-353.

THE CRAB NEBULA AND PULSAR originally appeared as chapter three of *The Emerging Universe,* edited by W. C. Saslaw and K. C. Jacobs, published by University of Virginia Press, Charlottesville, 1975.

THE GREATEST SUPERNOVA SINCE KEPLER was originally published as "1987A: The Greatest Supernova Since Kepler" in *Reviews of Modern Physics,* volume 60 (1988),pages 859-871. The author is grateful to colleagues working on SN 1987A who were generous with preprints, reprints, advice, and pictures, particularly to Larissa N.Alexeyeva, W. David Arnett, John N. Bahcall, Gregory Benford, Sidney A. Bludman, David Branch, David F. Chernoff, Roger A. Chevalier, Edward L. Chupp, J. Cooperstein, I.John Danzinger, Eli Dwek, Michael W. Feast, James E. Felten, Claes Fransson, Wolfgang Hillebrandt, Peter Höflich, Sidney H. C. Kahana, Margarita Karovska, Robert P. Kirshner, Leon Lucy, F. Curtis Michel, Luis A. Milone, Ken'ichi Nomoto, Keith A. Olive, Franco Pacini, Nino Panagia, David M. Rank, Fredrick Reines, David N. Schramm, Noriako Shibazaki, James M. Truran, Michael S. Turner, Robert V. Wagoner, J. Craig Wheeler, James Wilson, and Stanford E. Woosley.

SO YOU WANT TO FIND A SUPERNOVA? is scheduled to appear in *Robotic Observatories,* edited by Sallie L. Balliunas, Fairborn Press, 1992.

REMNANTS OF HISTORICAL SUPERNOVA was originally published as "The Crab Nebula and Other Historical Supernova" by Virginia Trimble and David H. Clark in the *Bulletin of the Astronomical Society of India,* volume 13 (1985),pages 117-126. The authors are indebted to the Raman Research Institute, Bangalore, and its director, Professor V. Radhakrishnan, for their truly princely hospitality during the conference at which the topics in this article were discussed. V.T. is also grateful to Professor J. C. Bhattacharyya and Dr. Vinod Krishan for the opportunity to visit the Indian Institute of Astrophysics, Bangalore, and to the Smithsonian Institution for travel support under its special foreign currency program.

GRAVITATIONAL RADIATION AND THE BINARY PULSAR originally appeared in *Nature,* volume 297 (1982), pages 357-358. Reprinted by permission. Copyright 1982 by Macmillan Magazines Ltd.

MASADA, SUICIDE, AND HALAKHAH originally appeared in *Conservative Judaism,* volume 33 (1977), pages 45-55. The author wishes to thank Meyer Greenberg and Evelyn L. Greenberg for their advice and counsel.

GRAVITATIONAL RADIATION DETECTORS was originally published as "Gravity Waves: A Progress Report" in *Sky & Telescope,* volume 74 (1987), pages 364-369.

THE BEST IS YET TO BE originally appeared in *Nature,* volume 303 (1983), page 382. Reprinted by permission. Copyright 1983 by Macmillan Magazines Ltd.

INFORMATION EXPLOSION was originally published as "A Controllable Aspect of the Information Explosion" in *Nature,* volume 310 (1984), pages 542. Reprinted by permission. Copyright 1984 by Macmillan Magazines Ltd. A fuller version appears in *Publications of the Astronomical Society of the Pacific,* volume 96 (1984), page 1007, under the title "Postwar Growth in the Lengths of Astronomical and Other Scientific Papers."

PROGRESS IS NOT OUR MOST IMPORTANT PRODUCT originally appeared in *Nature,* volume 302 (1983), page 20. Reprinted by permission. Copyright 1983 Macmillan Magazines Ltd.

DEATH COMES AS THE END was originally published as "Death Comes as the End: Effects of Cessation of Personal Influence upon Rates of Citation of Astronomical Papers" in *Czechoslovak Journal of Physics,* B36 (1986), pages 175-179. The author is indebted to Professor L. Woltjer for suggesting the hypothesis that originally motivated this investigation.

YOUNG VERSUS ESTABLISHED AMERICAN ASTRONOMERS was originally published as "Some Characteristics of Young Versus Established American Astronomers" in *Publications of the Astronomical Society of the Pacific,* volume 100 (1989), pages 646-650. The author is indebted to Helmut Abt for his usual generous and thorough job of refereeing. This investigation was partially prompted by discussions with Martin Rees.

PRESTIGIOUS START was originally published as "For Unto Him that Hath . . ." in *Nature,* volume 342 (1989), page 11. Reprinted by permission. Copyright 1989 by Macmillan Magazines Ltd.

SELF-CITATION RATES IN ASTRONOMICAL PAPERS was originally published as "A Note on Self-Citation Rates in Astronomical Papers," in *Publications of the Astronomical Society of the Pacific,* volume 98 (1988), pages 1347-1348.

BRIEF ENCOUNTER WITH A FACILITIES MANUAL was a contribution to Adrian Mellott's *Stupid Paper Project,* which was never published. It appears here for the first time.

ASTRONOMICAL CONFERENCES is a merging of a couple of short articles written for *IAU Today,* the daily newspaper of the International Astronomical Union 20th (Baltimore) General Assembly, edited by Stephen P. Maran.

TRIMBLE'S LAWS was a response to a request from Andrew Fraknoi of the Astronomical Society of the Pacific for something to put in a special issue of *Mercury* on Women in Astronomy. It appears here for the first time.

MARTIN J. REES was originally two articles: The Formation of Martin J. Rees" in *IAU Today,* 8 August 1988, and "Rees Pieces Together How Galaxies Form" in *IAU Today,* 10 August 1988. *IAU Today* was the daily newspaper of the 20th (Baltimore) General Assembly of the International Astronomical Union. It was edited by Stephen P. Maran and published by the IAU.

MAARTEN SCHMIDT was originally two articles: "Maarten Schmidt and the Rise and Fall of Quasars" in *IAU Today,* 5 August 1988, and "The Short, Significant Life of Quasars" in *IAU Today,* 8 August 1988. *IAU Today* was the daily newspaper of the 20th (Baltimore) General Assembly of the International Astronomical Union. It was edited by Stephen P. Maran and published by the IAU.

BEATRICE M. TINSLEY was originally published as "Extracts from *My Daughter Beatrice*" in *Journal of College Science Teaching,* November 1987. *My Daughter Beatrice* is a biography of Beatrice M. Tinsley written by her father, Edward Hill, and published by the American Physical Society in 1986.

FRITZ ZWICKY will be published as "Zwicky, Fritz (1898-1974)," a biographical entry in *The Encyclopedia of Cosmology,* edited by N. S. Hetherington, Garland Press.

Index

Concepts and Classes

accretion disks, 140
anthropic principle, 81, 94
atmospheres, 76, 110

background radiation, 26, 58
binary stars, 75, 108, 126, 137, 190, 207, 225
black holes, 30, 86
Brans-Dicke theory, 209
brown dwarfs, 86

cataclysmic variables, 137, 208
Cepheid variables, 68
Chandrasekhar limit, 136, 150, 152, 190, 208
citation rates, 244
clusters of galaxies, 62, 85, 88
comets, 9
constants of nature, 79, 81
continental drift, 100
cosmic rays, 295
cosmology, 49, 286

dark matter, 21, 84, 227, 296
deuterium, 91
DNA, 96
dolphins, 101
Dyson spheres, 106

early universe, 54, 95
electromagnetic radiation, 224
evolution, 102

Friedmann models, 63

galactic evolution, 52, 123, 124, 191, 283
galaxies, 52, 64, 87, 122
general relativity, 224
glitches, 163, 201
gravitational lenses, 22, 296
gravitational radiation, xv, 207, 222

GUTs (Grand Unified Theories), 89, 98

Halakhah, 210
HR diagrams, 68, 69

inflation, 90
interstellar dust, 50
interstellar molecules, 77, 99

Jewish Law, 210

large number hypothesis, 82, 97

magnetic fields, 152
meteor showers, 16

n-body simulations, 22
neutrinos, 170
neutron stars, 30, 74, 136, 162, 164, 226, 295
novae, 15, 133, 152
nucleosynthesis, 26, 70, 74, 91, 119, 174, 176, 178

origin of life, 77, 99

planetary conjunctions, 9
planetary nebulae, 150, 132
planets, 76, 99, 108
plate tectonics, 100
pulsars, 136, 159, 164, 184, 201, 207, 226

quasars, 28, 44, 165, 279, 282

Roche lobes, 128, 129, 130, 131, 135, 139, 140, 142
rotation curves, 87
RR Lyrae variables, 69

Seyfert galaxies, 31
star formation, 65, 127
stars, 96

steady state cosmology, 23
stellar structures & evolution, xiii, 67, 149,
    160, 168
suicide, 210
superclusters, 24, 295
supergalaxy, 23
supernova remnants, 158, 163, 191, 195
supernovae, 16, 75, 121, 136, 152, 158, 167,
    187, 195, 226, 294, 295
SUSY/SUGR, 89, 98

Texas Symposia, 28, 289

water, 96
white dwarfs.30, 74, 136, 137, 138, 147, 190,
    226
Wolf-Rayet stars, 168

X-ray sources, 132, 178, 183

People

Abell, George Ogden, 20
Abt, Helmut A., xvi, 268, 238, 235
Adams, Walter S., 26, 148
Alpher, Ralph, 26, 32, 120
Ambartsumyan, V.A., 22
Ångström, Anders, 84
Applegate, James, 91
Argyle, Edward, 110
Armour, Richard, 117
Arnett, W. David, xiv, 67, 172

Baade, Walter, 30, 31, 75, 116, 200, 201, 295
Babcock, Horace W., 20
Badawy, Alexandre M., xi
Bahcall, John N., 157, 201
Bath, Geoffrey, 144
Baym, Gordon, 31
Becker, Robert, xvi, 203
Begelman, Mitchell, 280
Bell, S. Jocelyn, 201
Bergh, Sidney van den, xiv, 192, 193, 201,
    203
Bergmann, Peter, 20, 28, 35
Bertelli, Gianpaolo, 151
Bessel, Friedrich Wilhelm, 84, 148
Bethe, Hans A., 26, 120
Blandford, Roger D., 280
Bohr, Neils, 31
Bondi, Hermann, 225, 288

Bonnor, William B., 25
Bracewell, Ronald, 107, 111
Brahe, Tycho 196, 199
Bressan, A.I., 151
Burbidge, Geoffrey R., 120, 201
Burbridge, E. Margaret, 120, 201, 291

Calvin, Melvin, 77
Cameron, Alastair G.W., 120, 201
Chadwick, James, 295
Chandrasekhar, S., xiv, 148, 200
Chiosi, Cesare, 151
Christy, Robert F., 201
Chwolson, O.D., 23, 32, 296
Clark, Alvan G., 85, 148
Clark, David H., 195
Clayton, Donald D., 51
Colgate, Stirling A., 201
Comte, August de la, 84
Cox, Arthur N., 153
Curtis, Huber D., 295

Darwin, Charles, 99
Demarque, Pierre, 108
Dicke, Robert H., 27, 44, 209
Dirac, Paul A.M., 82
Doroshkevich, Andrei G., 27
Drever, Ronald, 231
Dyck, Melvin, 127
Dyson, Freeman, 20, 29, 30

Eddington, Arthur S., 32, 82
Eggen, Olin J., 24, 123
Einstein, Albert, 20, 31, 32, 224, 296
Ekers, Ronald, 50
Evans, Robert 188, 192, 194

Fang Lizhi, 38
Farmer, Roy Stanley ("Uncle Roy"), xii
Feinberg, Gerald, 109
Feyman, Richard P., xv, 29, 82
Filippenko, Alexei V., 189
Flamsteed, John, 192
Forward, Robert L., 230
Foucault, León, 84
Fowler, William A., 28, 35, 120, 154, 201
Franhofer, Josef von, 84
Friedman, Herbert, 201
Frost, Edwin B., 23, 296
Frost, Robert, 74
Ftaclas, Christ, 242

Gamow, George, 20, 255, 26, 32, 44, 120, 200
Gaposchkin, Cecilia Payne, 276
Ginzburg, Vitaly L., 31, 35
Gott, J. Richard, 110
Graves, Robert, 14
Green, Richard, 154
Greenberg, Meyer, xv
Greenstein, Jesse Leonard, xiv, 148, 153, 284
Gunn, James E., 20, 46, 50, 201, 284

Hardorp, Johannes, 117
Harrington, J. Patrick, 108
Harrington, Robert, 108
Harris, William E., 238
Hart, Michael, 107, 109, 110
Hayakawa, Satio, 38
Hayashi, Chushiro, 26, 32
Hazard, Cyril, 284
Herman, Robert, 26, 32, 120
Hertz, Heinrich,, 221
Herzog, Emil, 24
Hewish, Anthony, 201
Hewitt, Paul, 284
Hill, Edward O., 285
Hill, Jean Morton, 286
Hirakawa, Hiromasa, 227
Hodge, Paul, 116
Hogan, Craig 91, 280
Holmberg, Eric, 21
Horace (Quintus Flaccus), 250
Hoyle, Fred, 24, 28, 31, 35, 49, 120, 201, 279, 293
Hubble, Edwin P., 52, 20, 115, 200
Huggins, William, 84
Hughes, David, 9

Iben, Icko Jr., 67
Infeld, Leopold, 225
Israel, Werner, 30, 31

Jeans, James H., 20
Johnson, Harold L., 267
Jones, Eric, 107
Josephus, 217

Kapteyn, Jacob C., 20
Kepler, Johannes, 17, 167, 196, 199
King, Ivan, 157
Kirchoff, Gustav, 84
Kirshner, Robert P., 174, 181, 201, 203
Koshiba, Masatoshi, 38
Kraft, Robert P., 137, 139

Kron, Gerard P., 116
Kuiper, Gerard, 116
Kumar, Shiv 108, 110

Landau, Lev 31, 202, 295
Lassell, William, 202
Lenard, P., 31
Levy, Marvin David, 210
Liebert, James, 156
Lifshitz, Evgenii M., 25
Lodge, Oliver, 32
Lundmark, Knut, 295
Lynden-Bell, Donald, 25, 123, 279

Maanen, Adriaan van, 202
Maimonides, 213
Mallama, Anthony D., 144
Mandl, R.S., 23, 31, 296
Maran, Stephen, xvi
Matthews, Thomas A., 284
Mayle, Ronald, 194
McClure, Robert D., 194
McCrea, William H., xvi, 20, 200
McKellar, Andrew, 26
McVittie, George C., 32, 200
Michaud, Georges, 154, 155
Middleditch, John, 203
Miller, Stanley, 76
Milne, Edward A., 20, 295
Minkowski, Rudolph, 152, 200, 283
Münch, Guido, xiv

Neugebauer, Paul, 7
Newcomb, Simon, 194
Newman, Ezra T. ("Teddy"), 225
Newton, Isaac, 56
Nomoto, Ken'ichi, xiv
Novikov, Igor D., 27

O'Connell, Robert F., 153
Oberg, James, 109
Oda, Minoru, 38
Ohm, E.A., 27
Oke, J. Beverley, xiv, 50, 148, 284
Oort, Jan H., 20, 85, 282
Osaki, Yoji, 144
Ostriker, Jeremiah P., 50, 201

Pacini, Franco, 42, 201
Paczyński, Bohdan, xiv, 67, 144
Page, Thornton Leigh, xiv, 23
Palmer, Patrick, 106, 111

Panagia, Nino, 176
Papagiannis, Michael, 106, 109
Penrose, Roger, 225
Penzias, Arno 27, 120
Peterson, Bruce A., 46
Phinney, E. Sterl, 280
Ponnamperuma, Cyril, 105, 110
Pringle, James E., 280

Rashi, 210
Ratnatunga, Kavan, 192
Rees, Martin John, xiii, xvi, 52, 91, 201,
    227
Reeves, Hubert 273
Reynolds, Stephen P., 203
Robinson, Ivor 28, 35
Rösner, W. 153
Rosenfeld, Leon, 31, 202, 296
Rubin, Vera Cooper, 20, 24
Russell, Dale xiii
Russell, Henry Norris, 148

Sachs, Rainer, 225
Sakharov, Andrei D., 25
Sandage, Allan 25, 123
Saslaw, William C., xiii
Sato, Katsuhito, 38
Sayers, Dorothy L., xiv
Schaeffer, Robert, 106
Schatzman, Evry, 154, 155, 200
Scheider, D.P., 284
Schild, Alfred, 28, 35
Schmidt, Maarten, xiv, 28, 44, 282
Schucking, Engelbert, 29, 35
Sciama, Dennis W., 279
Sciatti, Hjalmar, 273
Secchi, Angel, 84
Seward, Fredrick D., 202
Shapiro, Maurice M., 25
Shapiro, Robert, 109
Shapley, Harlow, 23
Sherrer, Philip, 91
Shipman, Harry L., 148
Shklovsky, Iosip (Josef) S., 31, 35, 168,
    200, 201
Singer, Clifford, 111
Sitter, Willem de, 20
Smith, Harlan J., 290
Smith, Sinclair, 22, 296
Soldner, J., 31
Srinivasan, G., 203
Stebbins, Joel, 116

Steigman, Gary, 51
Stokes, George, 84
Strömberg, Gustaf, 23
Struble, Michell, 242
Struve, Otto, 267

Tammann, Gustav A., 192
Tarter, Jill, 109, 111
Taylor, Joseph H., 202, 207, 226
Tey, Josephine, 14
Tinsley, Beatrice Muriel Hill, xiv, 50, 123,
    285
Tinsley, Brian, 289
Tremaine, Scott D., 50
Truman, Harry S., 294
Tworog, Bruce, 108

Uomoto, Alan, 174
Urey, Harold, 76

Vaucouleurs, Gerard de, 24

Watkins, Vernon, 105
Weaver, Thomas A., 174
Weber, Joe, xv, 20, 44, 162, 221
Wegner, Gary, 155
Weisberg, Joel M., 207
Weiss, Rainer, 231
Weizsäcker, C.F. von, 25
Westphal, James, 157
Wheeler, J. Craig, xiv, 175
Whipple, Fred, 18, 200
Whitman, Walt, 194
Wickramasinghe, N.C., 50
Wild, Paul, 24
Wilson, James, 194
Wilson, Olin C., 20, 200, 296
Wilson, Robert, 27, 120
Winget, Donald E., 154
Woltjer, Lodewijk, xiv, xvi, 31, 201
Woosley, Stanford E., xiv, 49, 169, 174
Wunner, Guenter, 153

Yockney, Hubert, 110

Zel'dovich, Yakov B., 20, 27, 35, 201
Zuckerman, Benjamin, 106, 111
Zwicky, Fritz, xvi, 20, 30, 31, 75, 85, 86,
    90, 160, 164, 188, 189, 200, 294
Zworykin, Vladimir K., 23, 31, 296

Objects, Places & Institutions

Algol, 126, 128, 129

Alpha Draconis, 5

American Astronomical Society, 38, 39, 246, 251, 282, 291

American Men & Women of Science, 251

American Physical Society, 39, 285

Andromeda Nebula (M31), 21, 88, 187

Annie J. Cannon Prize, 292

Annual Reviews of Astronomy & Astrophysics, 40, 41, 42, 258

Antares, 70

Apollo 17, 228

Argonne National Lab, 222

Astronomical Journal, 235

Astronomy & Astrophysics Abstracts, 41, 44, 45, 199, 261

Astronomy & Astrophysics Survey Committee, 261

Astrophysical Journal , 291, 265, 235

AXAF, 178

Babylonia, 15

Baltimore MD, 273

Beta Lyrae, 128

Betelgeuse, 70

Bethlehem, Star of , xii, 9

Cas A, 192, 198

Cheops' pyramid, xi, 3

China, 196

Coma Cluster, 21, 90, 296

Crab Nebula, xiv, 29, 32, 75, 86, 158, 192, 199, 227, 243, 295

Cumberland Gap, 104

Dallas TX, 289

Egypt, 3

Einstein Observatory, 183

Galileo mission, 228

GINGA, 178

Glarus (Switzerland), 294

Greenbelt MD, 276

Hubble constant, 62, 115, 116

Hughes Research Labs, 230

Hyades, 150, 115, 116

Inst. of Astronomy (Cambridge), 279

International Astronomical Union, xvi, 38, 40, 42, 167, 245, 250, 265, 273

Io, 103

IRAS (InfraRed Astronomy Satellite), 127

IUE (International Ultraviolet Explorer),186

Jahresbericht, 41, 199

Jericho, 94

Jerusalem, 11

Jupiter, 9, 109

Local Group, 61, 88

Lydda, 214

M 31 (Andromeda Nebula), 21, 88, 187

Magellanic Clouds, 122, 167, 202

Mariner 6 and 7, 228

Mars, 17, 102

Maryland, Univ. of, xv, 222

Masada, 211

Mercury, 207, 224

Milky Way Galaxy, 85, 115, 192

MIR (space station), 178

Mira Ceti, 197

National Science Foundation, 291

NP 0532, 163, 164, 165, 204

Orion, 6, 66

Palomar Observatory Sky Survey (POSS), 242

Pasadena CA, 276

Pioneer 10 and 11, 228

Pleiades, 150

Polaris, 5

Procyon, 148

Project Orion, 107

PSR 1913 + 16, 207

Publications of the Astronomical Society of the Pacific, 235, 258

ROSAT, 178

Royal Astronomical Society, 282

Saturn, 9

Science Citation Index, 236, 244

Sirius, 148

SNR 0540 –69.3, 204

Socorro NM, 276

Solar Maximum Mission (SMM), 179

Stenonychosaurus, xiii

Sun, 115, 117

Supernova 1987A, 124, 167, 189, 192

3C 273, 165
TTauri, 127

Ulysses (Solar Polar), 228
United Kingdom, 252
Universe, 60

V 471 Tauri, 129, 141, 142
Venus, 10, 109, 197
Virgo Cluster, 22, 58, 296
Voyager mission, 103

WIMPI (words in mean paper index), 238

Yale University, 292

About the Author

Virginia Trimble holds appointments at the University of California, Irvine, where she is Professor of Physics, and at the University of Maryland, College Park, where she is Visiting Professor of Astronomy, and her husband, physicist Joseph Weber, is Professor of Physics emeritus.

Born and raised in California, Trimble received her B.A. (magna cum laude) in Astronomy and Physics from UCLA in 1964, then went on to the California Institute of Technology, earning an M.A. in Astronomy and Physics in 1965, and a Ph.D. in Astronomy in 1968. In her postdoctoral years, Trimble taught at Smith College, served as a consultant to the Hughes Research Laboratory in Malibu, California, and spent two years as a Visiting Fellow at the Institute of Theoretical Astronomy in Cambridge, England. She earned an M.A. from Cambridge University in 1969.

Professor Trimble's research interests encompass a wide variety of astronomical topics; she writes and lectures on, for example, white dwarfs, supernovae, pulsars, black holes, binary stars, galactic evolution, gravitational radiation, and the sociology of astronomy. She has published more than 150 research papers and articles, and is regularly invited to deliver papers at astronomical society conferences and meetings. She has traveled throughout North America and to Europe, China, India, and Japan to speak at a variety of astronomy and astrophysics symposia. In 1986, Trimble received a National Academy of Sciences Award for Scientific Reviewing.

In addition to her research and teaching duties, Trimble is an active member of some dozen astronomical societies, including the American Astronomical Society, the Royal Astronomical Society, the European Physical Society, and the International Astronomical Union. She is a fellow of the American Association for the Advancement of Science and of the American Physical Society. Her advisory and organizational activities are myriad: she has chaired the International Astronomical Union Working Group on Supernovae, the Astronomy Section of the American Association for the Advancement of Science, and one of the panels of the Astronomy and Astrophysics Survey Committee. She is the editor of *Comments on Astrophysics* and associate editor of the *Astrophysical Journal.*

QB
51
.T75
1992

QB
51
.T75

1992

24.95